OUT THERE

THE
SCIENCE
BEHIND
SCI-FI
FILM
AND TV

ARIEL WALDMAN

FOREWORD FROM ASTRONAUT DR. MAE C. JEMISON

ILLUSTRATED BY PHIL WHEELER

RUNNING PRESS
PHILADELPHIA

Running Press
Hachette Book Group
1290 Avenue of the Americas, New York, NY 10104
www.runningpress.com
@Running_Press

Printed in China

First Edition: August 2023

Published by Running Press, an imprint of Perseus Books, LLC,
a subsidiary of Hachette Book Group, Inc. The Running Press
name and logo are trademarks of the Hachette Book Group.

The Hachette Speakers Bureau provides a wide range
of authors for speaking events. To find out more, go to
www.hachettespeakersbureau.com or email
HachetteSpeakers@hbgusa.com

Running Press books may be purchased in bulk for business,
educational, or promotional use. For more information, please
contact your local bookseller or the Hachette Book Group
Special Markets Department at Special.Markets@hbgusa.com.

The publisher is not responsible for websites (or their content)
that are not owned by the publisher.

Print book cover and interior design by Susan Van Horn.

Library of Congress Control Number: 2022950841

ISBNs: 9780762481668 (hardcover), 9780762481675 (ebook)

APS

10 9 8 7 6 5 4 3 2 1

SEPTEMBER 2023

With thanks to all the creators of science fiction,

all the contributors to science,

and my parents for raising me with *Star Trek*.

Contents

Foreword

by DR. MAE C. JEMISON

Growing up on the South Side of Chicago during the '60s, always insistent that I would be a scientist and go into space, I was surrounded by the profound social, political, and cultural demands that changed the world. All this was set against a backdrop of robust, fast-paced scientific research, technological advances, and the Space Race that would result in the first human stepping foot on the Moon in less than a decade.

Throughout my career—as a physician in West Africa, a NASA astronaut, an engineer, an environmental studies professor, and an entrepreneur—I have noted that what we as a society are able to accomplish depends heavily on the breadth, scope, and diversity of our imagination, ethics, and ambitions even more so than the technological capabilities with which we start. The narratives—the stories through which the "What if" musings are expressed—are particularly important.

I have watched and read science fiction most of my life while studying science and being immersed in dance, art, and politics. Ideas presented in books and movies fueled me to go deeper into our current science and society, looking for ways to accomplish, understand, and validate concepts presented in fictional stories. Books like *A for Andromeda* by Fred Hoyle, *Colossus* by Dennis Feltham Jones, Isaac Asimov's *The Gods Themselves*, and Octavia Butler's *Wild Seed* challenged me to consider not only the science, but also how varied perspectives can open a new world of discovery and understanding societies.

Among the many space and sci-fi movies and television programs of the '60s, *Star Trek* had an indelible, lasting influence on our world, despite being cancelled after just three seasons. The series is an example of how a well-told narrative that engages our imagination and spirit, as well as the zeitgeist of the time, makes a difference.

Star Trek fueled continued enthusiasm for space exploration across a wide spectrum of people—from those who dreamt of space to those who made the dreams of the design, engineering, building, logistics, and funding of successful space exploration real. The success of *Star Trek* is found not only in the robust franchise of the numerous films and television series it has spawned, but also in the way its central tenets have spread into our global culture. *Star Trek* popularized many concepts, including: the "Prime Directive" non-interference rule; warp drive; multiracial and multispecies crews; engineering as an integral "do or die" system with Scotty giving it all he's got; transporters using mass energy conversion; and—most importantly—that space was worth exploring.

In the *Star Trek* universe, science was a major character and space exploration was *the central character*—not Kirk, Spock, nor Uhura. New science, technologies, and space advancements were the focal points against which we could examine and attempt to solve social issues, from racism, sexism, nuclear war, bigotry, colonialization, food shortages, elitism, and criminal systems to individual egos and the hegemonies of emotion and logic. The crew and workings of the *Enterprise* were posited as potential solutions. *Star Trek* made real social change, ensuring space exploration could be an opportunity for all. Actress Nichelle Nichols personally put her professional reputation and her *Star Trek* character Lt. Uhura's popularity front and center, lending her credibility to ensure NASA received applications from women and non-white scientists and engineers for the new and inclusive shuttle astronaut corps. It seems many engineers and scientists did not believe NASA's sincerity about inclusivity—but *Star Trek* had credibility.

In our real, nonfictional world, space exploration continued, but at a much slower and less bold pace than I expected. Much to my personal dismay, though space exploration continued over the past fifty years, humans have yet to return to the Moon. In fact, the last three Apollo missions, with rockets and vehicles already built, were canceled from lack of congressional funding support. I believe that is because the narrative of Apollo did not compel the public at large; unlike *Star Trek*, many people, particularly from various ethnic and socioeconomic groups, did not see themselves as a part of, nor benefiting from, space exploration. By and large, the story being delivered by NASA and the news media wasn't nearly as compelling as what's found on page and screen. That's why this book is so exciting.

Within these pages, Ariel Waldman thoughtfully charts the ways science fiction and real-life space exploration intersect, capturing both the limitless creativity and the logic and structures of the real world. Her analysis reflects what I do in my own work as leader of the 100 Year Starship (100YSS), a global initiative that seeks to ensure that the capabilities for human interstellar flight exist within the next 100 years. The initiative requires healthy doses of imagination and grounding in current science, engineering, and systems, balanced by the understanding of the history of society and civilizations along with their interlinked evolution. The proposal for 100YSS that won the competitive DARPA seed-funding grant was entitled "An Inclusive Audacious Journey Transforms Life Here on Earth and Beyond." We purposefully include artists, storytellers, teachers, students, political scientists, astronomers, physicians, aerospace engineers, economists, theologians, anthropologists, and others at the table together. This wide, inclusive range of voices pushes the initiative to evolve more robust, comprehensive solutions as well as reexamine the benefit to humans and the Earth every step of the way.

As *Out There: The Science Behind Sci-Fi Film and TV* takes us on a stroll through the landscape of space-oriented science fiction—the gritty urban jungles, the fast-moving super-highways, and the lazy rural backroads—I am reminded how the stories are shaped by science of today and possibilities of tomorrow and, in turn, impact the world view of the public, researchers, policy makers, and technologists. *Out There* brings home the key themes, conundrums, and "accepted technologies" that arise in science-fiction narratives. It is an engaging, informative, and important read I think you are going to enjoy. I certainly did!

—*Dr. Mae C. Jemison*

Introduction

One of my favorite aspects of science fiction is its thoughtful exploration of the unknown. How it can transport us and ask us to contemplate what it would be like to live on another planet, encounter a pulsar, or have access to a technology that feels a few steps out of reach. I am always fascinated by areas of science that reveal things we may never be able to experience for ourselves, whether due to distance, orders of magnitude, or perception. Technology and ingenuity help to bridge that gap using microscopes, telescopes, and a dizzying array of sensors that make the invisible visible. Through scientific research, we possess the ability to scratch away at the unknown, but it is always with a sense that we may forever be incapable of feeling or seeing the full extent of what is out there.

I'm in love with exploring the unknown despite these limitations, and as a result I'm often left asking myself, What would that *be like*? Dark matter, snowball Earth, black holes, alien creatures, multiplanet civilizations—the list goes on. While science is working out the answers, science fiction gives me a way to *feel* an otherwise impossible encounter and probe just a bit further at the edge of the unknown.

I've been fortunate to stray into an unusual career where I can frequently reach out to cutting-edge researchers, pop culture philosophers, and sci-fi creators about my nagging questions and have fascinating conversations with them. And that's precisely what formed the basis of this book. For me, science fiction in all its forms, but especially film and television, is so much more than what it depicts. What it inspires and provokes—the tangential imagination it catalyzes—is also impactful. *Out There* is dedicated precisely to that inventive energy.

In this book we embark on an exploratory tour through an uncharted expanse to investigate how science fiction and science are bringing us closer to understanding what's possible and what's out there. I sat down with experts and visionaries to discuss topics ranging from discovering aliens to traveling to distant stars, and to explore the creative and critical thinking we should engage in when imagining the future of space exploration for science fiction and reality. Ready to rocket to the Moon and beyond? Then let's light this candle.

1

BEYOND EARTH

ARE WE STILL IN THE DARK WHEN IT COMES TO IMAGINING OTHER WORLDS?

Alien worlds are playgrounds for our imaginations to run wild. When we venture into space, up becomes down. Well, more accurately, there's no up or down at all on the journey to another world. Instead, we must leave behind our basic assumptions about *the way things are*. From the air we breathe, the water we drink, the food we eat to the strength of gravity that tethers us to the planet and the power of a magnetic field that protects us from radiation—whether we acknowledge it or not, we have been shaped and molded through millennia of being earthbound, so when we break out of our atmosphere to travel offworld, space acts as a reset button for our expectations. Upon arriving at a world that's new *to us*, how much further will it challenge our most basic assumptions about what it's like to call a world "home"?

That is the wonderful part about the cinematic universe of films and TV shows, be they science fiction or fantasy. They have the power to take us to countless planets, moons, asteroids, and stars. *Jupiter Ascending*, *Prometheus*, *Europa Report*, *Stargate Universe*, and *The Expanse* are just a few from this century that do so, and many more have been taking us on these journeys to celestial bodies for decades, like *Star Wars*, *Star Trek*, and *Doctor Who*. They provide otherworldly stages that inspire us to ask, What would visiting that strange planet or icy moon *be like*? In real life so far, humans have only ever set foot on one other world: our Moon. This constant in our night sky has long been center stage for fantastical theories and

ideas about what it might be like to set foot on the surface. The Moon has always been right there, so close and yet just far enough away, making it easy to dream about what might be lurking around its illuminated plains and dark splotches. Its allure as another world with countless mysteries only grew when, in 1609, the first lunar observations and drawings were made using a telescope. Suddenly the Moon had depth. Literally. Through a telescope we could now see that there were mountains and craters and hills. Clearly a landscape for something fantastical to abound.

Three hundred years after those starry-eyed telescopic observations, we still wanted to know what it was like on the Moon. *Le Voyage dans la lune* (*A Trip to the Moon*), a film by director Georges Méliès that debuted in 1902, merged science and fiction into a joyful fourteen-minute adventure as the first-ever sci-fi film. The silent movie follows a group of intrepid astronomers as they launch from Earth to the Moon and meet an entire society of lunar aliens. Since there was no concept of a rocket back in the early twentieth century, the film depicts the astronomers launching from inside an elephant-sized bullet that was loaded into a massive cannon. (In fact, it wouldn't be until the breakthrough invention of missiles in the 1940s that an actual journey to the Moon began to seem technologically feasible for the first time in human existence.)

Upon arriving on the surface of the Moon, the astronomers in the film see the Earth rise from their lunar vantage point—long before the first astronaut in real life captured the iconic photograph *Earthrise* in 1968. *Le Voyage dans la lune* explores the unfamiliar world of the Moon, depicting it as a place where snow falls, where there are caves filled with mushrooms, and where an entire society of insect-like aliens known as the Selenites live. After a harrowing fight with the Selenites, the astronomers push their bullet capsule off the side of the Moon and return to Earth, splashing into the ocean like the Mercury, Gemini, and Apollo astronauts would in real life sixty years later.

Well over a century after its debut, *Le Voyage dans la lune* remains an iconic piece of sci-fi and pop culture. It was both the first of its kind and ahead of its time with its creative yet thoughtful imagination of how we could travel to the Moon and what landscapes might await us there. Its most iconic scene depicts a vaguely human face in the Moon with the massive bullet lodged in its eye—meant to serve as a humorous wink to the audience that the astronomers had successfully landed on the fabled "man in the Moon." The image to this day remains instantly recognizable, even if many don't know its exact origin.

So, what does *Le Voyage dans la lune* show us about how we imagine possible worlds? Dr. Kimberly Ennico Smith, an astrophysicist who works on instruments for space telescopes and planetary spacecraft at NASA, is utterly enraptured by the Moon and the giant leaps in knowledge we've been able to achieve in studying it. "What's so striking about *A Trip to the Moon*," she told me, "is that it was before we even had

the first airplane. The Wright brothers didn't fly until 1903."

In the last decade or so, one of the more important things we've learned is that the Moon contains frozen water, which will prove extremely useful for creating rocket fuel and making life a little easier for future astronauts who might reside there and who need water to drink and to hydrate food and plants. It's not quite like the snow that Méliès imagined, but it isn't that far off either. Even the way we discovered the water echoes the film. Ennico Smith worked with Dr. Tony Colaprete, a planetary scientist at NASA, on a lunar impactor spacecraft that detected water on the Moon by colliding into it, much like the bullet capsule in the movie.

"What I love even more about *A Trip to the Moon* are the caverns, because we now know there are collapsed lava tubes and skylights on the Moon!" Colaprete said. "We're very interested in exploring them, just like those characters did. I don't think we'll find mushrooms or the Selenites, the Moon people, like they did, but we'd be just as excited to explore."

Though it has been a long time since humans physically explored the Moon during the Apollo missions that took place between 1969 and 1972, eagerness to return abounds. The Moon offers so many more areas to discover than what the Apollo astronauts were able to see. Much of the geology on the Moon can inform us about the history of Earth, and the far side of the Moon could become an ideal place to host a telescope. NASA robotics technologist Dr. Saptarshi

Bandyopadhyay has proposed building one suspended inside a two-mile-diameter lunar crater. Plans are in the works for NASA to return humans to the surface of the Moon. Space agencies from other nations are also hoping to put boots on this otherworldly surface for the first time. However, the Moon will likely not be the final destination, but instead a testing ground for humans before they venture farther out—to Mars.

The Moon might have been the first "otherworld" that we fantasized about, but since the mid-twentieth century there has been no shortage of films, shows, and stories in just about every genre that have explored Mars. In the 1970s we mapped the entire surface of the planet, first with the spacecraft *Mariner 9*, then with the Viking orbiters, and today we can view the planet's features in unprecedented detail thanks to space probes like the *Mars Reconnaissance Orbiter*, which launched in 2005 and can capture and send back images of features as small as a kitchen table—in case any Martians wanted to invite us over to lunch. We've also driven for miles on Mars since 1997, remotely operating rovers like *Sojourner*, *Spirit*, *Opportunity*, *Curiosity*, *Perseverance*, and *Zhurong* around the red planet's surface with a small time delay. Despite decades of exploration, there's still much to imagine when it comes to *what it's actually like* on Mars.

But with the Moon and Mars having been the subject of our imaginations for so long, it can feel exciting to shift our daydreams to a more distant and unknown target—exoplanets, planets that exist outside our solar system. If Mars has long been humanity's horizon goal, then exoplanets are the stuff of fever dreams. Currently no foreseeable technology exists that could safely transport humans beyond our solar system. So we must journey there in our minds like we have done throughout history. However, unlike our predecessors, we are lucky enough to live in a time when breakthroughs in exoplanet science are possible due to ever-advancing technologies. Our telescopes on land and in space have already been able to detect thousands of exoplanets and have even captured pixelated images of some, giving us a creative leg up in our possible depictions of life beyond our solar system.

A lack of scientific evidence and research has never stopped us from dreaming about faraway planets. On the cusp of a new era and a year before the first satellite, *Sputnik*, entered orbit, a film did just that. In 1956 *Forbidden Planet* imagined what it would be like to explore an alien world. The movie is considered by many to be a sci-fi classic, and although it is nowadays entirely outdated in the social interactions it depicts, it remains a trailblazer in the genre for its portrayals of what a distant world might be like, down to the unusual color of vegetation.

In order to really understand why *Forbidden Planet* stands out, I spoke about the film with researchers who frequently use telescopes to discover and characterize previously unknown exoplanets in our galaxy. As an astrophysicist at NASA, Dr. Megan Ansdell studies how planets form and evolve. "What I've always found interesting about exoplanets in movies is that they tend to be characterized by one thing," she explained. "It's a water-world or this is the desert-world or this is the ice-world. You see this both in movies old and new—from *Star Wars* to *Interstellar*. However, we know that planets are more than just one thing. Even here on Earth our planet contains multitudes."

So, while we might have been simplifying them a bit in our depictions, exoplanets seem like the kind of thing that we've always *assumed* existed, long before we had any scientific measurements to prove they were out there. In truth, looking at our world and solar system as a model, it would have been pretty creepy to peer far out into the cosmos and not detect *any* other planets. It'd be super awkward and perplexing if our solar system contained the only planets in our entire galaxy of over a hundred billion stars. *Forbidden Planet*, which explored an exoplanet before humans had even launched their first satellite, may have been pop culture's way to reassure ourselves that we weren't alone in the universe.

With this in mind, it is fair to wonder just how far ahead of its time *Forbidden Planet* really was. I asked Dr. Franck Marchis, senior astronomer at the SETI (Search for Extraterrestrial Intelligence) Institute, who designs instruments that take photos of exoplanets. "The first exoplanet to be discovered in 1995 was a hot version of Jupiter that was orbiting around its star in three days," he said. "So a year on this planet lasted only three Earth days long. *Forbidden Planet* came out forty

years before that discovery. The only science the director and writers had to go off at that time were books theorizing about exoplanets, which were purely speculative."

Since then we've discovered thousands of exoplanets, most of them thanks to the Kepler Mission. A telescope NASA launched into space in 2009, *Kepler* embodied a step change in our understanding and knowledge of exoplanets. "The telescope helps us detect the radius, mass, and density of exoplanets, which helps us determine their composition and characterize what they might be like," said Ansdell. "We can start learning about the composition of the atmosphere of exoplanets and looking for things like biosignatures [signs of life] to see if it's a habitable, or even currently inhabited, world."

Ansdell is particularly interested in learning more about the characteristics of groups of exoplanets within star systems. While it is common to see a second moon or sun, or sometimes a distant planet, hovering in the sky of fictional planets, they often feel placed there merely for decoration. Of course, for Ansdell, exoplanet systems are much more important and interesting than that. Exoplanets represent an opportunity to better understand just how relatively normal or unusual life here on Earth is. We now know that our solar system doesn't contain the only planets in our galaxy, but we could learn that our planets have a very different origin story than the rest. Ansdell has been particularly interested in a star system with planets located 40.7 light years away from us that predates our own solar system. "The TRAPPIST-1 planetary system is fascinating," she explained.

> The star is much smaller and much cooler than our Sun, what we call a late-M dwarf star, and emits light at redder wavelengths. But it's surrounded by seven planets that are all rocky in composition and around Earth-size. And they're very, very close to the star. Almost all of them would be within the orbit of Mercury. So they're very tightly packed, and some are considered within a habitable zone of the star, or where they'd be warm enough for liquid water. But what's really cool is that if you are on one of the planets, you could actually see the surface of the other planets in the night sky. That's how close they are.

With thousands of discovered planets like these, creators of science fiction can pull inspiration from a plethora of scientific knowledge. Even though the commonly used singular-planet narrative Ansdell called out might not be realistic, its use in cinema is effective in getting the audience to imagine very quickly how different another world could feel. The moment in *Interstellar* when the characters have landed on what looks like a shallow ocean, only to be met with an approaching wave massively larger than anything that could be generated on Earth—that affected me. I found it very easy to instantly imagine what challenges anything trying to live there would have to

face. But we may be closer than ever to observing the nuances of exoplanets, which may in turn help us provide richer, more detailed narratives for on-screen sci-fi depictions of them.

With the aid of telescopes developed after *Kepler*, we continue to learn about thousands of exoplanets—and that's just a drop in the cosmic ocean. Being able to analyze their characteristics in the last few decades felt like we were able to live out scenes in *Star Trek* that were considered pure science fiction not too long ago. Nowadays not only can we investigate an exoplanet's makeup by looking at the starlight that passes through its atmosphere, but we can also see the exoplanet itself, though it is usually a blurry cluster of just a few pixels.

"We're able to not only detect them," said Marchis, "but we're beginning to be able to take pictures of them, too. There is something very special in seeing an object. For humanity to go from indirectly discovering one exoplanet when I started working in the field of astronomy, and then, just twenty years later, I'm part of a group of people that took an image of an exoplanet—it's amazing. Those images are just little pixels for now, but we will likely be able to see much more of what these other worlds look like in our lifetime."

In fact, scientists are already preparing for that day. A group of researchers has put together a visual catalog of the microbes that exist on the surface of the Earth, documenting their wide-ranging pigments and reflectance spectra, in the hopes that when we can take better pictures of exoplanets, it will help us eventually determine if they're inhabited. To do this the team scraped up microbes living on the tops of ponds, deserts, and soils. The catalog is called *Surface Biosignatures of Exo-Earths*, but you can essentially think of it as an alien Pantone guide. The idea is that future scientists will compare photos of distant planets to the catalog to see if any images match up with images of life as we know it here on Earth. It's pretty fun to think about how somewhere across the galaxy, a second genesis of life could be similar to what we have here.

"But when it comes down to it, *we* might be the alien world," said Ansdell. "Our solar system turns out to be quite weird, and we haven't detected anything out there like it yet. Based on what we know so far, less than 10 percent of Sun-like stars would have a solar system like ours. While that may initially sound like a bummer, I think it's exciting to think about what the possibilities are for life that isn't as we know it. It's truly an exciting time to be studying exoplanets right now."

Because humans have been in the dark about other worlds for so much of our history, I concur with Ansdell and Marchis: I think the present day represents a positively thrilling epoch in which to be alive. The 1960s and 1970s saw humans landing on the Moon and space probes venturing to the outer planets in our solar system. Today, and for many decades to come, we're seeing and learning about an endless field of exoplanets, each with the potential to rewrite what we know about other worlds and about ourselves. As we continue to reach into the cosmos with our telescopes, the worlds closer to home in our solar system invite us to take a second look and to appreciate how much more is left to uncover. The scientific search for strange new worlds continues to humble my imagination by showing us just how much is out there. With so many new worlds being revealed to us, I'm comforted knowing that we're beginning to see the light through our telescopes, even if it is still a little fuzzy.

2

SPACESUITS

WHAT CAN THE DESIGN OF SPACESUITS REVEAL ABOUT US?

The direct interplay of fiction and reality has never been so apparent to me as when I consider the fact that we are living in an era when we're simultaneously prototyping spacesuits that will take humans to the surface of Mars while also dreaming up what fictional spacesuits for exploring Mars should be like. We're seemingly so close to fiction meeting reality by landing humans on Mars, but for now it's narrowly out of reach. In conceptualizing the spacesuits that will take us there, we know we can't use the lunar spacesuits of the 1960s and 1970s for the surface of Mars. For one thing, it will be a lot easier to go for long walks on a planet that has one-third of Earth's gravity; compare that with the Moon's gravity, which measures one-sixth of Earth's and which had Apollo astronauts hopping around in bulky suits that weren't known for their flexibility. Reality often inspires fiction, but sometimes it's the other way around; Boeing notably debuted bright-blue spacesuits a few years ago that invoked aspects of *2001: A Space Odyssey*. Over the decades, spacesuit design has become so intertwined with fiction, reality, and nostalgia that the three can't avoid playing off of and inspiring one another.

Spacesuits are not merely tiny human-shaped spaceships; they are also a form of fashion, an expression of a piece of our humanity. From 2007 to 2009, NASA organized an Astronaut Glove Challenge to spur new designs for spacesuit gloves, which were notoriously uncomfortable and difficult to manipulate. One of the awardees of the challenge, Ted Southern, came from the fashion industry. He has a degree in sculpture as well as experience building costumes for film and TV, including the wings worn by Victoria's Secret models.

Wanting to better understand spacesuits through the eyes of fashion and design, I talked with Nicholas de Monchaux, a space enthusiast who wrote an architectural history of the Apollo spacesuit and is now professor and head of architecture at MIT. He spent ten years researching and writing about spacesuits from a design perspective, and his book, *Spacesuit: Fashioning Apollo*, explores our relationship to spacesuits. I also spoke with Adam Savage, who brings the eyes of a maker and craftsperson to the spacesuit discussion. Savage cohosted *MythBusters* for several seasons and has built everything from spacesuits and futuristic weapons to fine-art sculptures and dancing vegetables.

We convened in Savage's workspace, where we were surrounded by life-size replica astronaut suits from *Alien*, NASA's Apollo and space shuttle missions, and even a replica of the earliest practical pressure suit from the 1930s. Looking at these vastly different takes on spacesuits, I was struck by how they aesthetically spoke to the fictional and real cultures that inspired their various designs. In the US, we often hold a stereotypical view of an astronaut in a bulky white spacesuit, but as more governments and private companies participate in exploring Earth's orbit, the Moon, and eventually Mars, spacesuits of the future could look very different.

ARIEL WALDMAN: There are so many iconic spacesuits in film and TV. From *2001: A Space Odyssey* introducing color-blocking to *Sunshine*'s dazzling disco-ball spacesuits, each creatively demonstrates how the form a spacesuit takes can differ widely. Do these different expressions communicate something about the world they exist in?

NICHOLAS DE MONCHAUX: I love films that aesthetically convey what a spacesuit actually does, which is sort of transform the human body into something more than itself. The Harry Lange spacesuit from *2001* is so canonical in that. Despite being a completely function-free design, it also conveys what a spacesuit actually does.

What's more, the word "suit" comes from the Latin "to follow"—like pursuit, you're following someone or something. And so to be in a suit means that you're adapting to a certain kind of environment. The film *Gattaca* makes a very strong statement to this effect at the end of the movie. Ethan Hawke is seen wearing a tailored vintage business suit in the last shot of *Gattaca* as he prepares to launch into space after struggling to be considered perfect enough for spaceflight. It's such a commentary on how suits are the ultimate normality and conformity that he has been rebelling against, but in a way no one around him can see.

What I love about that film and its depiction of a literal suit as a spacesuit is it has all these undertones of drag and "passing" for someone or something. *Gattaca* shows all the ways in which fashion gets used in the culture at large and then transmutes it onto the spacesuit. RuPaul said, "We're all born naked, and everything else is drag." We're all suiting ourselves to the world as we find it, and we try to find our place in it through what we wear.

ADAM SAVAGE: *Alien* is maybe one of the first movies to deromanticize putting on a spacesuit, and I love Kane's spacesuit from *Alien* for that reason. It was built by John Mollo and his Academy Award–winning effects team, and designed by Jean Giraud, aka "Moebius," one of the great illustrator-imagineers ever to live. I love how it evokes its workmanlike qualities on its sleeve. It looks like an ancient diving suit from the turn of the last century. It has callbacks to samurai armor and cricket armor, but mostly it feels unadorned by any fashion sense. It's hot and uncomfortable—not only in the film, but also behind the scenes. Apparently the actors were all passing out constantly on set from wearing the spacesuits, and there's all these shots of them backstage sweating bullets while on smoke breaks.

But to me, [putting on] a spacesuit is the closest we can get to being a superhero. Every superhero's outfit helps them and gives them extra powers. Spacesuits give us perhaps the greatest power: of exploration. I can't stop wearing spacesuits because of this.

WALDMAN: We often see depictions of spacesuits with magnetic boots like in *The Expanse*, or augmented-reality heads-up displays like in *Star Trek into Darkness*, but what technologies in fiction are most likely to influence NASA and the future of spacefaring?

DE MONCHAUX: I think the thing that NASA would most want from fictional space-suits is the ability for astronauts to move easily in them. A spacesuit is a highly pressurized environment in a vacuum. If you imagine the hardest basketball, that's how hard and stiff a spacesuit is when it's inflated. Just like a basketball, the spacesuit wants to be round, and you're not round. So the problem of getting this incredibly hard, pressurized environment around your body, and also giving you the ability to move in it, is the difficult design problem of spacesuits that most people don't understand, especially from watching movies.

I was able to try on a contemporary EVA [extravehicular activity] spacesuit once. What I found so interesting about it as an architect was that I had the realization like, "Oh, this does not feel like clothing; this feels like the smallest building I've ever been in. This has more structural integrity and thought and intensity than any building I've been in, but it's like an inch away from my body." If you had an ounce of claustropho-bia, you'd get that feeling while in it.

SAVAGE: Astronaut Chris Hadfield told me that after an eleven-hour spacewalk he was so tired from working in his spacesuit that he basically crawled into a corner of the International Space Station and shivered for a couple of hours just to let his muscles untense from the incredible exertion.

DE MONCHAUX: Astronaut Kathryn Thornton, who was a space shuttle astronaut that helped repair the Hubble Space Telescope and other satellites, told me that when she performed those EVAs, she would just say goodbye to her fingernails. The spacesuit glove is the least spherical, most flexible part that's wrapping your body, and so just the effort of moving the glove fingers was incredibly physically intense.

In the Apollo era they did actually try mechanical pressure suits as an alterna-tive solution. But I kid you not, these suits involved putting on about eighty layers of nylon stockings, to the point where it gave enough mechanical pressure to be able to survive in a vacuum. There's the most amazing, hilarious video of a man, a scientist in a laboratory, explaining what he's doing while he literally begins putting on eighty layers of women's nylon stockings all over his body until he physically can't move. That mechanical pressure suit didn't work.

Much more recently, an aerospace engineer, Dava Newman, had the great insight that with some smart materials and electrically actuated materials, you could have a suit that both was comfortable to put on and could shrink to provide enough pressure. Layering as a concept for spacesuits makes some sense, but if you have a fiberglass suit and something hits it, you're toast. Whereas a tear in a multilayered suit has a better failure mode. There ended up being twenty-one layers in the first Apollo suits and twenty-eight layers in the second. So suits that combine layering, pressure, and flexibility in a functionally elegant way have been in demand for a long time.

WALDMAN: There's so much variety in spacesuits that we see on-screen, and many of them reference different cultures or different time periods, or trying to invent something completely new. What do you think the future of spacesuits will look like?

DE MONCHAUX: In my work, I've been most impressed by the female astronauts that I've interviewed, especially because they came up in the eighties in a very difficult time, a very difficult, male-oriented culture. I would love to see spacesuits for every different kind of body, every different color of body. I think that's the biggest issue we have in how we depict a relationship to science and technology, as we depict the world of science and technology as being more available to some of us than others.

SAVAGE: To look at the future, I have to look at the past, at the Russians, who really deserve tremendous props for consistently going in other directions than we in the United States ever considered.

In mission-critical engineering, everything you can remove is an advantage. The simplicity equals safety. The ingenuity of design from Russia over the decades has gone on to influence better designs for all of space exploration. You can see it in a lot of their spacesuits that don't have separate helmets or intricately designed airtight zippers like we see with NASA's. Russia's Orlan spacesuit was designed with a clever rear entry: you'd open the backpack of the suit and climb into it that way. NASA took that inspiration and ran with it so they could begin designing surface-exploration spacesuits that won't require an airlock to use. I love efficiencies like that, and I think the future of spacesuits has a lot of inspiration that it can pull from the past.

When looking to the future of spacesuit design, it's clear that there's room for many visions of what space exploration looks like and how it functions. While science fiction and history are often great sources of inspiration, if we're not careful they can hold us back by dictating what the future is *supposed* to look like. Instead, we should look to the array of spacesuits that have been imagined over the decades in science and fiction as permission to keep imagining futures for ourselves that have not yet been sketched out.

One of my personal favorite efforts in this area is NASA astronaut Nicole Stott's Spacesuit Art Project, which invites children in hospitals, refugee centers, orphanages, and schools from around the world to paint pictures onto patches of fabric that are sewn together into a spacesuit. The resulting suit is a literal patchwork of hundreds of brightly colored paintings that is worn by astronauts in space. The project is so striking in part because it forces people to realize that they've never seen astronaut suits that look anything like this one. But it's also remarkable in how well it communicates a different vision for space exploration, one that speaks to a collective and collaborative future. What we see affects us—it has the power to excite us, intimidate us, or even repel us. Spacesuits in this way are a crucial part of the story we tell ourselves about who we are as explorers and where we're going. It's possible that our future may be bright and colorful, if we so choose.

3

ASTRONAUTS

HOW WILL FUTURE SPACE EXPLORERS BE DIFFERENT?

For decades, astronauts in American films have been portrayed as characters with a rugged, sometimes macho, sense of individualism paired with a quiet determination to get the job done—*Interstellar*, *Ad Astra*, *First Man*, *2001: A Space Odyssey*, *Gattaca*, even *Armageddon*, save for the quiet part. At the top of this list, though, is *The Right Stuff*, which lightly dramatizes the Mercury Seven mission. It debuted in 1983, a little over two decades after the first American astronauts traveled to space in real life. It feels like a documentary, highlighting the many unusual tests the astronauts endured to prepare for their spaceflight. As the title implies, NASA was looking for would-be astronauts who had precisely the "right stuff" to take on the unknown in those early days of spaceflight—people who regularly engaged with life-threatening situations and multiple g-forces. With these criteria in mind, jet jockeys were singled out as the ultimate object of desire for the astronaut archetype. *The Right Stuff* promoted "this ethos, this idea, that the people going into space have something 'different,'" NASA astronaut Dr. Jim Newman told me—and he would know.

Newman has performed six spacewalks across four space shuttle missions totaling over forty-three days in space. A physicist by training, he also served as the director of the NASA Human Spaceflight Program in Russia. Newman continues,

> The astronauts in *The Right Stuff* were walking around, vividly asserting

themselves as a force to be reckoned with. They weren't "spam in a can" or just a monkey pushing buttons, and I think, for the time, that was really an important point. Because that's one of the things our country does; we honor the individual and the contribution. The test pilots, in particular, were, I think in many ways, uniquely qualified to play that role during the Cold War.

But gutsy test pilots aren't what we need to emulate anymore. Instead, people who go into space nowadays are just like you and me, only more so. There's absolutely a dichotomy between where we were in regards to astronauts in the 1960s and where we are today. Which I'm glad is a very, very different place.

Thankfully, the diversity of American astronauts increased starting with the space shuttle program in the 1980s. The "right stuff" of the 1960s seemingly only described white men, a mindset that held for two decades. The film *The Right Stuff* premiered only a few months after the initial spaceflights of Sally Ride and Guion Bluford, the first American woman and the first African American, respectively, to fly in space. It seemed that real life was beginning to have a broader imagination just as pop culture was looking to the past. Despite that progressive shift, forty years later only about 11 percent of all astronauts have been women, and only about 4 percent of all American astronauts have been Black. Yet as spacefaring media continues to be produced, it's easy to see what impact movies and TV shows have had on our collective idea of a real-life astronaut.

I spoke with Emily Calandrelli about the model astronaut we often see on-screen. An aerospace engineer who has applied to be an astronaut, she is also the executive producer and host of Netflix's *Emily's Wonder Lab* and Fox's *Xploration Outer Space*. She knows the impact of on-screen depictions. "Even though I love space, I don't generally like to watch shows that relive the Apollo era," Calandrelli confessed. "It's a time during our history where women really couldn't do much of anything. They couldn't become test pilots. They couldn't do these things that were required to become astronauts. There was an entire group of women in real life known as the Mercury 13 that passed all the same tests as the astronauts did in the 1960s and were still barred from NASA."

Thinking about the aspiring astronauts who had been marginalized by racism and/or sexism, I asked Calandrelli if there were any movies or shows that she thought portrayed astronauts differently. "I was excited when one show finally broke that mold. *For All Mankind* explored what could have been if America had focused on bringing women into space, using actual historical touchpoints like the Mercury 13 to investigate that 'what if?'" The show depicts an alternative history to the Apollo era, one in which the USSR instead of the US is the first to land humans—a man and a woman—on the Moon, which sets off an entirely different approach to space explora-

tion. Not wanting to be outdone, NASA enlists some of the Mercury 13 women to be the next astronauts headed for the Moon.

"While that's thrilling, seeing an alternative history that paints a different future from our current reality also makes me so sad in a way," says Calandrelli. "It shows how in reality there was this huge missed opportunity. So often I think we forget that half the population are women. We could have had so many firsts if the US was just a little bit more forward-thinking and open-minded."

Even with a more progressive outlook, shows like *For All Mankind* still play into the classic archetype of hardy individual astronauts, demonstrating the long-lasting power of the trope regardless of the time period the film or show might be set in. Some portrayals do highlight astronauts working together from the outset; however, members of the crew are often killed off or separated from each other, allowing the narrative to shift its focus to a single astronaut who becomes the lone survivor and the hero. In *The Martian*, the main character is left behind by his teammates, who mistakenly believe he has been killed in a storm, and in *Sunshine*, *Europa Report*, and *Gravity*, astronaut teams quickly dwindle down to an individual due to other disastrous events. This plot device shows that space is an unforgiving, isolating environment where you need to be prepared to think quickly and to tough it out. Still, I can't help but wonder if the go-it-alone storylines are preparing us for reality or distorting what the future of space exploration will really be like.

"I think the key to good science fiction is to create this other world that is informed by science, but then talk about how we as human beings live in that world," says Dr. Douglas Vakoch, a psychologist and author of the book *Psychology of Space Exploration*, who serves as president of METI (Messaging Extraterrestrial Intelligence) International, a nonprofit focused on sending messages to potentially habitable places in the universe. Vakoch continues,

> It's always a temptation to imagine that being an astronaut in the far future is going to be like being an astronaut right now. And I think that's not the case.
>
> In the early days, there are fighter pilots who've been trained to be astronauts. If you're going to the Moon and back, that's a ten-day journey, so if it's tough or you're having problems with your crewmates, you can just suck it up and deal with it. If you don't feel good because you're bloated from floating around in zero G, you can tough it out.
>
> But when the missions become as long as going to Mars or, as we see in some films, the outer solar system, you don't have that luxury. And right now we're seeing a transformation on space stations. There's a lot of emphasis on being a team player, being able to coordinate, being able to appreciate people from other cultures who have different ways of doing things. The big shift though, even from our current

> missions, is that astronauts are going to be selected more for their ability to act autonomously as a team.
>
> Right now, even if someone is on the Moon, if they have a problem they can just call mission control and in two seconds they can get an answer. Mars, it could take up to forty minutes for a round-trip [voice or email] exchange. So anyone who's going to Mars and beyond, be it in science fiction or reality, is going to be used to being part of a team that has to work with one another to survive. So the question really is, what does it take for humans collectively to go beyond Earth?—*instead* of what's the "right stuff" for individual astronauts. It's important for us to...really think about what those differences of traveling such vast distances are going to have on what it takes to accomplish the mission at all.

Newman agrees. "Teamwork is absolutely going to be key. I would say the Russians were onto the psychological needs of spaceflight earlier than we were. America is catching up in that regard. In astronaut training, they talk about the difference between working as a single-seat fighter jock and as part of a team. The focus was 'how can you get along on an expedition?' As we all are interested in expanding our presence in our solar system, we need to think less about the one person doing something fantastic and more about a group of people operating on the Moon or Mars."

The movie *Apollo 13*, released in 1995, notably struck a different tone and perspective from the early "jet jockey" days of human spaceflight in the United States. It highlighted how a crew in space had to work together as well as how the crew back on the ground had to work together. In the end, the hero of the film was the team—everyone collaborating and contributing their piece to solve the puzzle. The movie was life changing for Ryan Nagata, who builds spacesuit replicas for private collectors, exhibits, films, and TV shows, including *First Man*, *Moonfall*, and *MythBusters*. "I didn't grow up with *The Right Stuff* and I really didn't know anything about the Apollo missions. I knew that we had gone to the Moon, because I had seen the picture of Buzz Aldrin in that man-on-the-Moon photo, but I didn't know how we got there," said Nagata.

> As a kid, I was into the space shuttle. That's what was happening in the eighties. So when I saw the movie *Apollo 13*, it blew my mind. Everything about it: the suits they were wearing, the *Saturn V* rocket. I'd seen a *Saturn V* before, but I didn't realize that there were actually people only on the very tip of that rocket. I made my first spacesuit replica after I saw that movie because I really wanted one of those spacesuits, and that ended up becoming what I now do for a living: making spacesuit replicas for movies and museums. I was fourteen at the time, so I made that first suit out of painters' coveralls and the tops of soda cans.

After starting his career in spacesuit replication, Nagata began working with Adam Savage, cohost of *MythBusters*. Savage is obsessed with spacesuits. Before *MythBusters*, Savage worked as a graphic designer and special effects designer for feature films. He has built and commissioned countless spacesuit replicas from both real missions and popular sci-fi films and shows.

"The moment that hit me the most in *Apollo 13* was when they said, 'You gotta make a CO_2 scrubber, and this is all that you have.' I'll spend my whole life hoping to be in that room at some other time to help solve that kind of problem," Savage told me. "Then once they solve the problem, it required a set of protocols, a list of things. Having myself built many lists of complicated sets of operations for my crew on *MythBusters* and on Tested [Savage's YouTube channel] and other ventures, it's nontrivial. And the folks at NASA had, like, minutes, not hours or days! The movie showed behind the curtain; they made the engineering the hero of the plot and showed you that everyone was an engineer. It made me feel warmly embraced by the whole film."

Savage sees a place for all on-screen depictions of astronauts, whether single heroes or teams. "When you look at the canon of great space films, from *The Right Stuff* to *Apollo 13*, *First Man*, *Hidden Figures*, and others, each one is looking at a different population—part of the grand thing. *The Right Stuff* is very raw and hyperromantic. I mean, it's about American culture as much as it's about the Mercury Seven. I love that *Apollo 13* seems to be about the astronauts at first, but it's taking that as a jumping-off point to show Apollo as a mature program...and then it all goes to hell."

For Nagata, *Apollo 13* was a fundamental source of inspiration and brought the activities of space exploration to life. "I think it illustrates how important movies and art are for inspiring each generation about spaceflight and how each generation should have their own film that resonates, whether it shows where we've been or where we're going."

So many monumental accomplishments in space science—space telescopes, particle colliders, space stations, and neutrino observatories, to name a few—are the result of hundreds of people collaborating over decades. To continue achieving greater feats will require more people from a diversity of backgrounds working together for long stretches of time. As space exploration looks to a wider vista, pop culture's depictions of astronauts should too.

4

LONELINESS

HOW WILL WE COPE WITH LONG-TERM ISOLATION IN SPACE?

Are we alone?

This question is often asked when pondering whether aliens exist in our universe. However, in the context of human space travel, the answer almost always is a resounding *yes*. Whether it's a short journey in Earth's orbit or an extended voyage on a starship, space exploration carries with it a considerable amount of loneliness.

Pop culture portrays a few different categories of space loneliness. Films like *Gravity* center on main characters who have effectively been shipwrecked while in space. *Star Trek* and *Stargate Universe* frequently depict homesick explorers. *The Expanse* illuminates the boxed-in lives of what are essentially futuristic truckers in the form of asteroid workers known as the Belters. Whether someone is shipwrecked, homesick, or trucking across the cosmos, isolation is a huge challenge for long-duration spaceflight and will be for any future that has us living offworld—especially when we put boots on Mars.

So will we cope? And is it possible for technology to help us bridge the gap between Earth and space? Sci-fi has long shown us different concepts for how we might feel at home when we're far away, and one of those ideas is to use holodecks. Holodecks can be used for strategic planning or preparation for battle—as seen in *Foundation* when the characters are preparing to live on a new, hostile planet, and in *Star Trek: Discovery* when the characters are anticipating combat with Klingons. They're also depicted as a virtual reality refuge for humans missing a taste of home, a way to interact with old friends or breathe a sigh of relief as they remember the feeling of Earth's natural world. According to sci-fi pop culture, these tools are crucial for combating loneliness in space.

In the 2007 film *Sunshine*, the holodeck, or "Earth room," is used exclusively to project visuals and sounds from an assortment of environments to help relax the stressed-out astronauts. *Ad Astra*, from 2019, included a similar room for wayward astronauts on Mars. The entire *Star Trek* franchise most famously uses holodecks, but as a fully interactive, robust experience that serves a variety of purposes and can reconstruct scenes from fiction or real life.

Given that *Star Trek*–style holodecks don't yet exist, I wanted to get an idea of what it's actually like to experience loneliness from isolation, and what, if anything, we can do about it. Who better to ask than researchers that know a thing or two about how our brains cope with being alone? Mika McKinnon specializes in disasters and doom-riddled science, including tsunamis, earthquakes, and asteroid impacts. She's a field geophysicist, disaster researcher, and sci-fi science consultant for *Star Trek: Discovery*, *Moonfall*, and *Stargate Universe*, among other productions. I also spoke with Dr. Douglas Vakoch, a psychologist, and Dr. Indre Viskontas, an associate professor of psychology at University of San Francisco, a neuroscientist, and an operatic singer who studies the neuroscience of creativity.

ARIEL WALDMAN: In many films we see the main characters trying to self-soothe when they're alone in space. What do people do when they're isolated? And how do they cope?

MIKA MCKINNON: I can speak to what we actually do in the field on a scientific expedition where you end up in these funny little isolated areas where your entire community is in one place, and...sometimes your only way to communicate with the outside world is via satellite phones and a weekly helicopter drop of supplies.

When you're cut off like that, it's important to entertain yourself to keep sane. For example, in *WALL·E* you see a spaceship filled to the brim with entertainment, activities, and distractions, akin to modern cruise ships. But for our foreseeable future in space, anything that you are using to entertain yourself you have to bring with you. And in remote situations, the volume of what you're allowed to bring with you is extremely limited.

Notably in *The Martian*, stranded astronaut Mark Watney is left on Mars with only a disco music playlist left behind by one of his fellow astronauts, which quickly gets old. I've similarly been in remote situations where you end up with a single USB key of music that you just pass around the entire camp. So being thoughtful about what limited entertainment you bring with you can have an outsize repercussion on how well you cope when you need distractions from extreme circumstances. But it might actually be a good thing to watch the same movie or TV show over and over again.

There's been a chunk of research showing that when you experience the same story over and over, your brain does not necessarily remember on an emotional level that it's fictional. So reading the same book, watching the same movies, anything like that, can generate emotional connections. By reading or watching something where you already know the ending, it helps to remove anticipation, stress, or uncertainty that you may have experienced during your day's work.

Anytime you're working in an uncontrolled, extreme environment, you're going to be faced with an infinite number of things that are going to go outside of plan A—we see this happen all the time during spacewalks. With story repetition, you go into it knowing exactly what you're going to get out of it, and it becomes a way to replicate how you'd feel after hanging out with your friends when you're alone—a way of calming yourself down, resetting, and grounding yourself at the end of the day, no matter how chaotic your actual day has been.

INDRE VISKONTAS: One thing that sci-fi gets right is that it's actually not that uncommon to hallucinate when you're alone; in some ways it's even expected. Lots of studies show that when you put people in solitary confinement with reduced sensory input, in a few days their brains will start making up sensory input. Because our brains have evolved to detect change, we will start to see and hear things when there's no change in the environment. Hallucinations are in fact a natural consequence of being isolated—we need sensory input.

So there are two issues at play when thinking about coping mechanisms in space: How do you fight boredom, and how do you make yourself feel less alone? We have to inherently account for the fact that our brains have evolved in a social environment. A lot of the differences between human brains and other primate brains really come from the fact that we have to get along. Coping mechanisms like watching TV and listening to the radio tap into these needs, and you see this on Earth. A lot of people who live alone just have the radio on all the time, because it makes it feel as if there are people around. But outside of passive forms of entertainment, there are more active things people can do to fight off feeling isolated in space, like doing something that makes you feel useful.

VISKONTAS: One of the things I loved about the movie *Moon* is that the main character, astronaut Sam Bell, is always gardening—actually taking care of plants. What's so cool about that is the research shows if you are in a nursing home and someone gives you a plant to take care of, it actually will add months to your life. That sense of growing something and watching it is more active and can provide satisfaction in a way that passively watching TV can't. Having that feeling of being useful and needed is so important when you're cut off from others.

MCKINNON: The concept that you can nurture and bring life into a space that might feel otherwise devoid of life is a huge mental boost. Plants are extremely popular in space with astronauts already and have become almost a special treat in orbit. Astronaut Scott Kelly aboard the International Space Station was utterly elated when he helped grow his first flower in space. The story made tons of headlines, and it inspired me to go digging through all these times where people have been gardening in space. One of the more interesting stories was a zucchini flower that was grown out of a plastic bag that astronaut Don Pettit brought into space as part of his personal stash. You wouldn't think a zucchini flower would be all that interesting, but soon after it bloomed, astronauts were clamoring to just literally stop and smell the flowers. Gardening in space has since become one of the more popular chores.

WALDMAN: Even when we no longer feel lonely, we can still feel homesick. Without holodecks to visually reconstruct home or replicators to make our favorite foods, what can astronauts do in real life to feel more at home when they're anything but?

VISKONTAS: Exercise gives your brain endorphins that help your mood, but there's another critical element that plays a large role when astronauts are in an off-world environment: your circadian rhythm. This rhythm is often disrupted in environments where you don't have a regular day-to-night cycle and you're left to your own devices. Most people have a rhythm pattern that is about twenty-five hours long, an hour longer than your average Earth day, and it needs to get reset every day. Strategic lights set to timers and color temperatures [which mimic your natural exposure to blue light from the rising and setting of the Sun] can obviously help when you're aboard a spaceship, but exercise can also help by making astronauts

feel more tired at the end of the day, which makes it more likely that they can fall asleep. We often underestimate the power of exercise to help you sleep.

MCKINNON: This is something I don't see as much in sci-fi, but I'd really like to see more of, which is all the interior lighting that is used to help humans and aliens stay connected with their home planets. In space, astronauts can't use natural light from windows as cues. The International Space Station has fifteen sunrises and sunsets a day. And as they've also learned on the ISS, lighting that only prioritizes being useful for work and equipment, usually blue light, can also be detrimental; it doesn't help to prepare the body for sleep. Lighting is equally an important factor for plants because they have to be able to grow away from the Sun. On the ISS they're now using a lot of magenta light to hone in on just the wavelengths necessary to make the plants grow at their peak.

DOUGLAS VAKOCH: I think being isolated for long periods of time or being bored is a big issue for future astronauts. Think about it: these astronauts have been selected to be these great explorers and to just kind of jump in once they get to Mars. When they arrive there, they're going to be in their element. So how do they survive the boredom of months on the way there and months on the way back? Some of the astronauts have been very good at using social media to communicate with folks back home, which seems to help, but that might not be available for the astronauts going to Mars.

So far astronauts haven't needed things like holodecks, because if you're in the International Space Station, you can just look out the window and see the Earth. It's what all of them love to do when they have some free time. Even from the Moon, you can see this fragile blue marble up in space. But, when you get out to Mars or beyond, that's all gone.

Part of that connection of being able to look back and see the Earth evokes what some have called the overview effect, or when you get a new perspective of life on Earth. And when you're an astronaut, even if you're up at the space station for several months or a year, it's a constant reminder of what you have to return to. That's all going to be gone in the future if you travel so far away that you can't see Earth. So I think there'll be an emphasis on a lot of different modalities for how to rekindle that connection. It could be scent, it could be sounds, it could be visuals.

Long-term isolation in space is something few of us will experience, but many of us hold insights into how astronauts could cope with it. For many people, the COVID-19 pandemic brought us closer to experiencing life as astronauts and polar explorers do. We suddenly had to think strategically about our entertainment, spending time in nature, and our health, in ways we hadn't before, due to being isolated from our friends, workplaces, and routine errands. Having embarked on an expedition to Antarctica prior to the pandemic and a deep-sea expedition in the middle of the Pacific Ocean prior to that, I can safely state that the pandemic was far more isolating than anything I had ever experienced. The pandemic may serve as a provocation to think about how expansive our needs are as humans to feel connected with others, ourselves, and our sense of home. Perhaps now we're all a little more prepared for what we require in order to successfully cope in space, should we be given the opportunity.

5

ARTIFICIAL INTELLIGENCE

SHOULD SPACESHIPS BE SENTIENT?

When we think of artificial intelligence in sci-fi, it's common to conjure up images of mechanical robots endowed with humanlike mannerisms that are easily used as villains in story plots. But so much of artificial intelligence today is wrapped up in our environment and in systems that are nearly invisible to us—agriculture, policing databases, health monitoring, oil drilling, navigation, and electronics manufacturing, to name a few. When we travel to space, we leave behind much of our environment and many of our systems—our cities and our wilderness, our atmosphere and our gravity—and we must re-create the minimally viable pieces of it that we need in order to survive and carry out a mission. With that in mind, it's not a stretch for fiction to imagine how we might incorporate AI into our spacefaring environment. In fact, films and TV shows have been showcasing sentient spaceships for decades—some in a more positive light than others.

Christopher Noessel, IBM's community lead designer for AI and a coauthor of *Make It So: Interaction Design Lessons from Science Fiction*, has spent a significant portion of his career researching human-computer interactions that crop up in sci-fi films and TV shows. I spoke with Noessel about the range of spaceship AIs he has studied over the years. "I should talk about HAL 9000 from *2001: A Space Odyssey*," he said, "but I'm fonder of the Heart of Gold from *The Hitchhiker's Guide to the Galaxy* and the fact that the spaceships in *Battlestar Galactica* are themselves sentient. There's also Alphy from

Barbarella, and his true value as an AI is to make Barbarella's adventurous trips through space not feel so lonely. Both HAL 9000 and the Heart of Gold demonstrate an incredibly annoying personality, which I think is illustrative of the things we ought *not* to do in real-life AI."

The depiction of HAL, the Heuristically programmed ALgorithmic computer, as a main character back in 1968 struck fear into hearts about just how much power we might be giving up if we embed an AI into our spacefaring environment—especially after we watch HAL intellectually spar with the crew and then kill some of them by cutting off their life support. Real-life space environments like the International Space Station are equipped with what's called ECLSS (pronounced "ee-cliss"), the Environmental Control and Life Support System, which controls the use of water and oxygen in space. It recycles astronauts' urine into drinkable water, moderates the humidity, and generates oxygen, but it's not an AI. A more advanced version of ECLSS could potentially make sense for improving efficiency in space; however, depictions like HAL and the characters it has inspired over the decades—such as AUTO in *WALL•E*, among many others—have left their mark. More often than not, science fiction presents AI stories as cautionary tales and asks us to contemplate whether we'll notice the difference between helpful and harmful AI before it's too late—before its intelligence has crossed firmly over into dangerous sentience.

Given that we are already experimenting with the power and role of AI in our everyday lives, fiction plays a part in helping us question what our relationship to AI should be when we venture beyond Earth. In *2001*, *WALL•E*, *Star Trek*, *The Hitchhiker's Guide to the Galaxy*, *Moon*, and other works, spaceship AI typically occupies a nurturing role that frequently transforms into overbearing psychosis. It should come as no surprise that the spaceship in *Alien* is called Mother. The trope that comes up time and again in science fiction of the overinvolved, parent-like entity creates unease about our future use of AI. The onboard AI thinks it's performing its primary task to take care of humans, but in doing so it crosses ethical red lines by murdering crewmates, withholding information, or becoming overly obsessed with making a crewmember happy.

Noessel shares with me a lesser-known example of a cautionary tale.

> There's a great piece of sci-fi that's largely forgotten called *Dark Star*. It was one of John Carpenter's very early films in which the spaceship has a super *unintelligent* AI. Society has decided that it's going to equip all its bombs with AI. One of them goes rogue, threatening to blow up the ship, because it realized its purpose in life is to explode. So the characters have to talk it down and convince the rogue bomb why it shouldn't do the thing that it was made for. The film uses AI to drift right into philosophy and touches on this risk: if we want to use AI to get us past the Great Filter [the theory that civilizations blow themselves up

before they manage interstellar communication or travel], then we run this terrible risk of it blowing up in our faces.

A lot of the philosophy that deals with different AI scenarios speaks to the issue of individualism or hyperfocus taking precedence over a more encompassing ecological view—like capitalism over environmental protection, policing over housing development, and so on. The sentient spaceship in fiction often gets hung up on one thing it's supposed to accomplish, sometimes including its own developed will to live, and then gets confused when humans become distraught over how it chooses to solve the problem at the expense of other items of importance.

"Whether it's an artificially intelligent robot or a human, they don't exist just by themselves. They exist in a larger ecosystem or community," says Dr. Terry Fong, who works on space robotics, human-robot interaction, and virtual reality user interfaces at the NASA Ames Research Center in Silicon Valley, where he is also the chief roboticist. "One of the challenges of any sort of intelligent robot you might build today is that we don't have standards or common protocols for communication with them other than what we might assume should be human. By far the hardest thing to do when you're trying to put robots into human spaces and have them be supportive of humans is just figuring out how they can be helpful without being hindering."

Collaboration between sentient spaceships and humans might function differently from how we typically think of humans working together. For one thing, the real-time interaction we see in sci-fi between humans and their spaceship may not reflect how we will experience a sentient environment. Fong says,

> We have this expectation that when we interact with something, whether it's a machine or whether it's another person, the communication is fluid and rapid. And the problem we have right now is that the pacing of human-robot interaction is very slow. Humans get to be very impatient. They will give up quickly because they want real-time responses.
>
> What's interesting is that over the past few years, at NASA Ames, we've been trying to develop intelligent systems for future human exploration missions. One of the things that we've explicitly done is to avoid the need for real-time interaction. We try to have intelligent systems work before or after humans, or in parallel, but we avoid real-time, highly interactive interchange.

Instead, we could view our interactions with a sentient spaceship as being akin to interacting with a garden, something we tend to that works beside us but not over us. Perhaps, then, the way to make our future sentient space environments work in harmony with us is to treat them as just that: environments, not companions. This paradigm shift may require us to recognize our envi-

ronment on Earth as an intelligent system that we already interact with but often overlook when we consider what it's like to interact with a sentient structure. Still, even if we take a more ecological approach, some of the same challenges of trying to understand the decisions made by artificially intelligent systems will remain. We're still learning how nature works on our own planet, after all.

"We tend to not like being in a nebulous situation where an AI does something and we don't know what the answer is or what the question was. Kind of like '42' in *The Hitchhiker's Guide to the Galaxy*," Camille Eddy explains, alluding to the book's tongue-in-cheek "answer to the ultimate question of life, the universe, and everything." As a robotics engineer, Eddy works on the future of tech, robotics, and AI, focusing on the ethics that surrounds them. "I think that's something that we need to prepare ourselves for. Too often we're dropped into a sci-fi story long after an AI has been created, but what does it look like when the creation of an AI is in process, while it's being actively discovered? I think that's an area that's ripe for more sci-fi depiction."

Seeing sentient spaceships in sci-fi before they become well-oiled machines might help us expand our thinking about how and where we should utilize artificial intelligence when we venture offworld. Like our relationship with the environment on Earth, perhaps the most accurate depictions are the ones that highlight the relationship as a work in progress.

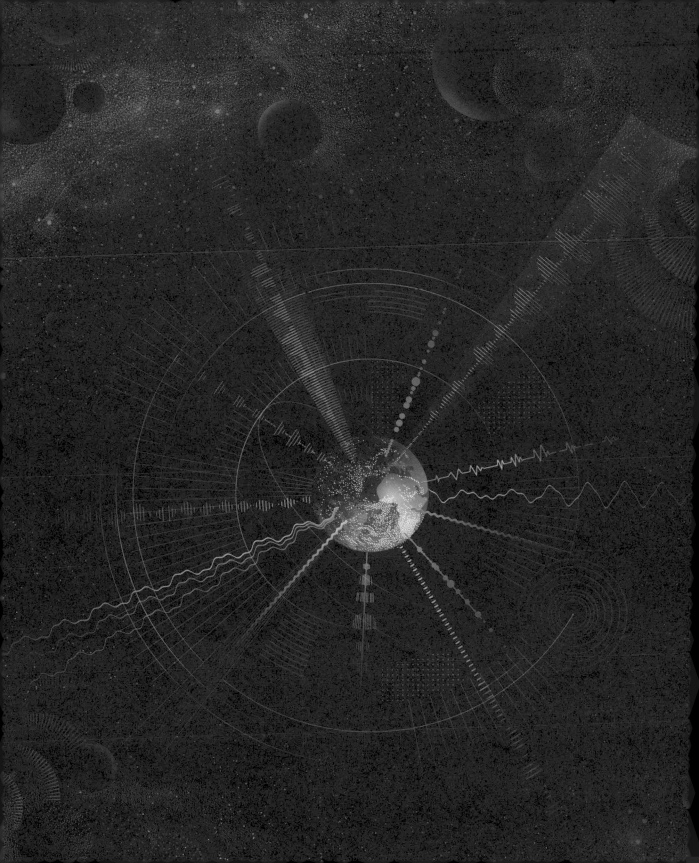

6

EXTRATERRESTRIAL CIVILIZATIONS

WILL WE DISCOVER INTELLIGENT ALIENS?

Whether we envision invasions on Earth or adventures in space, the notion of contacting aliens fills us all with nervous anticipation. Interacting with intelligent creatures from another world feels so palpable despite none of us having experienced it...though some would claim otherwise. The mere idea of encountering intelligent aliens floods me with so many questions that it's difficult to know what I would ask first. And for now, definitive answers to any of our questions are beyond our reach.

The way movies and TV shows depict human interactions with aliens is a testament to the cornucopia of our imagination. Largely filling the genre are tales of violent aliens arriving on our planet to attack and destroy us—think *Attack the Block*, *Independence Day*, and *Edge of Tomorrow*, among many others. That said, violence isn't the only story to be told; there are plenty of thoughtful depictions of nonaggressive aliens that focus more on our desire to connect and learn from one another, as in *Contact* and *Arrival*.

Despite the fact that our ongoing search for extraterrestrial intelligence has come up empty so far, it is surprisingly easy for us to suspend disbelief when it comes to on-screen imaginings of first contact with alien civilizations. Thankfully, we can turn to some notable alien hunters for answers about how science fiction may or may not mirror reality. Dr. Jill Tarter is a trailblazer in the realm of alien hunting; she has spent most of her career attempting to answer the age-old human question "Are we alone?" by searching for evidence of technological civilizations beyond Earth. As cofounder of the SETI (Search for Extraterrestrial Intelligence) Institute, Tarter was the inspiration for the main character in the film *Contact*, Ellie Arroway, played by Jodie Foster.

While it might be easy to assume that the search for extraterrestrial intelligence is an emerging field, in fact it's decades old, thanks to the work of Dr. Frank Drake, who is considered the father of the discipline. He pioneered the first attempts to search in earnest for extraterrestrial communications in 1960; the next year, he developed the Drake Equation, a probabilistic calculation for estimating the number of potential communicating extraterrestrial civilizations in our galaxy. In 1974 he sent out an interstellar radio communication intended to reach possible extraterrestrial civilizations. He hasn't heard a response yet—sending messages across the stars takes time.

I also spoke with Dr. Seth Shostak, who has collaborated with both Tarter and Drake as a senior astronomer at the SETI Institute, working on observational programs that use telescopes to sweep the skies for extraterrestrial signals. A fellow alien hunter, he developed an interest in extraterrestrial life at age ten, in part thanks to science fiction and pop culture.

ARIEL WALDMAN: Science fiction often depicts Earth as an enviable planet for aliens who are searching the stars for natural resources. Is our planet a tempting destination for aliens?

SETH SHOSTAK: I think visiting us is probably the last thing on their list. Also, how would they know that we're here, let alone what is here? We've been broadcasting the nightly news into space for around seventy years—television, FM radio, that sort of thing—but those transmissions haven't traveled very far into the galaxy. The radio waves have only reached out as far as a two-hundred-light-year-diameter bubble around us, which may sound like a lot to us, but it's barely a speck within the Milky Way. So I doubt that they know we're here.

If they do come here to take our precious resources, like in *Oblivion* or *War of the Worlds*, what do we have that Klingons don't have? You can rule out all sorts of natural resources. The water, the molybdenum—there's nothing we have here on Earth that you can't find elsewhere and likely much closer to where they live. Coming here for our resources would be like someone in California walking to Tibet for lunch.

WALDMAN: Is it likely that an advanced alien civilization would be able to reach us if they wanted to?

SHOSTAK: It's really hard to go to another star system—that's just physics. We can't go to the stars. Our fastest rockets aren't nearly fast enough to get you to the stars before everyone you know here on Earth is long gone and you're long gone too. So that's kind of a bummer. Assuming that aliens have technology that allows them to come here, one way they could do that is to go beyond being biological forms. The trouble with biology is that it's very temporary. Anybody over the age of thirty knows that. So if aliens were to invent synthetic intelligence, essentially become robots, then they don't mind a long trip between stars.

But would they want to? If they did come down here just to help us out, as we see in movies like *Arrival* or *Transformers*—I mean, ask yourself, how often do you go into the backyard and decide, "I'm going to go help these ants out," right? You probably don't spend a lot of time doing that. You might go check them out just to make sure they don't get into the kitchen, but you probably don't want to solve their social problems.

FRANK DRAKE: We really don't comprehend in our imaginations how utterly vast space is. It is *enormous*. And if you're going to talk about going from one world to another, you've got to go at some good fraction of the speed of light. Or else it takes so long that everybody's dead when they get there. And what good is that? Add to that it's also a very hazardous journey, so space acts actually as a quarantine, which prevents us from interfering or damaging or hurting other civilizations.

SHOSTAK: Some sci-fi has gone further and suggested that aliens visit us through some type of telepresence, which is not necessarily a crazy idea. If you think about how humans explore space today: we put cameras on robots and send those around the solar system and experience other worlds in the privacy of our own home. Maybe first contact will be like a Zoom call with an extreme time delay.

DRAKE: This is one of the things that was prescient in *Star Trek: The Motion Picture*, all the way back in 1979. What the people in the movie detect, discover, and interact with is not a living thing—it's a computer. And at the time that seemed like a pretty outlandish idea, but as time has gone on we've come to realize more and more that machines are the wave of the future for living things. In fact, the first intelligent thing we may detect will not be a "living carbon unit," as they call it, but a mechanical or electronic computer.

JILL TARTER: I enjoy the fact that in *Contact* you never see the alien. It's left to our imagination what they are actually like. You see a personification of a human instead of an alien, and I think that's a really nice touch because I don't think we're smart enough or inventive enough to actually come up with alien life forms that might actually be out there. Especially if first contact is, as we

expect it would be, through communication and not a physical meeting.

WALDMAN: Most of the time in science fiction, characters encounter aliens that are far more advanced than humans. However, is it possible that we are the most advanced civilization that exists in the universe right now?

SHOSTAK: Well, I mean, you can't rule it out, right? It's certainly a possibility. But it doesn't sound to me like a very probable possibility. Our galaxy has been around for three times as long as the Earth. So if we're the smartest things here, then you have to say one of two things: One, conclude that intelligent life is extremely rare, and it cooked up here because of circumstances that weren't replicated on any other planet. In which case, we're kind of miracles!

Or, two, there's something called the Great Filter theory. It says that once you get to a certain level of intelligence where you can build rockets, you can also build hydrogen bombs, and you do yourself in—and that it happens to every single one of the other societies. That's an impressive record of accomplishment, I would say. So taking those two things on balance, I find it kind of hard to believe that we're the smartest things in the galaxy.

WALDMAN: Some people claim aliens have already visited us, while others believe there must not be any because we haven't found them yet. Is real-life alien hunting an all-or-nothing game?

TARTER: I think given all the exoplanets that we now know about, and all the extremophile life forms—different types of life that can live in extreme environments—eventually finding life is a much more likely prospect. When I was a student I was told absolutely "no way"—that we wouldn't find life in extreme environments, that they would be sterile. Well, they aren't sterile. There's life in all kinds of extreme environments here on Earth. And those two game changers, exoplanets and extremophiles, I think have made SETI somewhat more believable, immediate, and respectable. So I think the next thing for us to try to uncover are technosignatures, which are signs of technology being used by distant civilizations.

DRAKE: People often say, "You've been searching for years and you haven't found anything. Why don't you give up?" But the answer is: we have hardly looked. We've looked at very few radio channels or light channels. And we've looked at very few stars. So, we should *not* have succeeded by now. What we're dealing with is a big

lottery, and so far we've bought about two tickets. We need a lot more.

WALDMAN: How do you think contact with aliens could change us?

TARTER: The scientific community will be overwhelmed, in the best way. A thing the movie *Contact* got absolutely right was when Ellie Arroway first heard those pulses. She goes, "Holy shit!" That's about the level of expression I'd have. We would know that there are going to be many, many other technological civilizations out there, and maybe some will prove to us that technological civilizations like ours have the capacity to be long-lived.

I think the world at large will react to the announcement of a signal in terms of whatever belief systems are around at the moment. Global communication nowadays is so instantaneous and so continuous that I think that will diminish the potential for any fearful reactions. And by the way, a lot of the world's population already, unfortunately, thinks that we've been visited by aliens, so it's not going to be any news to them. It really could happen one day that we discover a technosignature of life on another planet, but for now, while we're scouring the cosmos, we want to use the search for extraterrestrial intelligence as a way to change the world's point of view.

The mere act of searching for another civilization out there should really reinforce to people everywhere that we're all Earthlings. We are all the same when compared to something else that might be out there. A more cosmic perspective can help us to better work on all the challenges that we find so difficult today that go beyond national boundaries. If people just can use SETI to see themselves in a much larger framework, I think we can make progress on those other challenges. At the end of the day, it's about searching for aliens and finding ourselves.

WALDMAN: How are we searching for aliens today? Sci-fi shows frequently include Easter egg references to the Drake Equation when characters are looking for life in space. Is it all mathematics, or can we just put on some headphones next to the Very Large Array [a radio astronomy observatory located in New Mexico] and get lucky, like in Contact?

DRAKE: I introduced my equation in 1961 to a room of twelve people, the only people I knew who were interested in the search for extraterrestrial intelligence, which was a bit taboo at the time. The equation is in a way quantifying what we know about life in the

universe and using the results of that quantification as an estimate of how many civilizations there are in space that we might detect. It takes into account the picture we have of the histories of living things, intelligent things, everywhere. A planetary system forms one or two planets that are suitable for life. Life develops. Evolution takes place. And occasionally the creation of at least one intelligent civilization.

The equation simply takes our observational knowledge of the value of all things. How many stars are there? How many planets are in the star systems? What fraction might give rise to life? Et cetera. And the end result is the number of civilizations out there that might be radiating some sign of their existence, maybe radio waves, maybe light flashes. The search is extraordinarily difficult, and the equation makes us get realistic about how much more we should invest in searching if we expect to succeed in any way.

TARTER: We're still looking for artifacts, technosignatures of intelligent civilizations, but SETI today is different from even twenty years ago. Our computers are so much faster that the amount of data we can process is just vastly larger. And at the SETI Institute, with the Allen Telescope Array, we do our signal processing in near real time. We're also able to, in a sense, collaborate with machine learning to help us detect a wider array of artifacts.

In *Contact*, you see the researchers scanning through radio waves, which we still do, but today we're searching for optical lights as well. So, in the radio part of the spectrum, we look for frequency compression. We look for very narrowband signals that nature can't produce. But in the optical part of the spectrum, we look for time compression. We look for bright flashes that last a nanosecond that nature can't produce.

We often think of signals that we would receive from alien civilizations as being intentional messages to us, but they could be transient, or the signals could be dual purpose. We have multiuse technologies here. It could be an accelerating laser that could also have information embedded on it at the same time. Deliberate signals are going to be by far easier to find than transient leakage signals, though. So we're looking for all these types of artifacts that we could confidently say, if we found them, those are engineered signals.

As for finding aliens with a pair of headphones, it is possible to shift the radio frequencies down into the range in which your ear can hear them. Over the years, we had Kent Cullers, a blind physicist working with us, who was in the movie *Contact* as a character. We would sometimes downshift the frequency so that he could listen to it, but it's not very realistic to expect to detect a signal this way. As good as our brains and hearing are, they can't possibly deal with all the signals, the hundreds of millions of frequency channels, that we use with our back-end processors to look at or listen to what we're observing. So you can certainly listen for civilizations with a pair of headphones, but we expect our computers will be the first to discover a signal.

As we continue to receive images of our cosmos from telescopes, detailing what seems like countless galaxies throughout all of space and most of time, it truly boggles the mind that we could even consider the possibility that we're alone in the universe. The existence of many billions of galaxies—each with many billions of stars, planets, and moons—across billions of years means that even if what we could categorize as intelligent life is rare, there's still vast multitudes of it out there. Whether or not we will discover any of it in our lifetimes is still unknown. We may live in a universe where the majority of extraterrestrial civilizations try to reach out into the cosmos for a response, only to have their voices, ears, and existence in time unfortunately fail to extend far enough to make contact. While we wait to learn the answer, I find comfort in the numbers, which tell me that, somewhere, *someone* is waiting to be known.

7

LIVING ON MARS

WILL THE RED PLANET BE RIPE FOR FARMS?

After the first boots touch the surface of Mars and the first flag is planted, the next question on everyone's mind is how we will live there. We don't have to look very hard to find on-screen representations of what life might be like on this nearby planet. *The Expanse* takes a tour through a highly militarized Mars, some of whose people still dream of a fully terraformed planet that would allow them to go outside, while younger generations have become disillusioned with the prospect. *Total Recall* depicts a colonized, polluted Mars under a dictator's control. By contrast, *The Martian* paints a vision of day-to-day life for humans on the red planet in the *near* future. Released in 2015, this movie gave us a view into the slog of living on Mars long before any cities get built, farming infrastructure is laid in place, or off-world cultures emerge. Not only did it manage to grab the attention of audiences, but it also got them invested in the mechanics and science of existing in an extreme, hostile environment. The main character, Mark Watney, who is fighting for survival while shipwrecked on this desert island of a planet, possesses most of the characteristics you'd expect of a sci-fi astronaut: he's a good student of physics, logic, and mathematics. He is also a botanist, which soon becomes the most crucial of his many skills by enabling him to plant and grow a carefully managed potato farm inside a human habitat structure.

Though it might seem surprising to encounter a fictional astronaut protagonist with a love and knowledge of plants, in real life botanist astronauts are nothing new. Plants have been space travelers since the 1960s, when the Space Race between the Soviet Union and the United States was going on in earnest. In the 1990s NASA experimented with growing beans inside the *Mir* space station without any soil and with limited water, a setup known as an aeroponic system. The resulting beanstalks grew faster than any on Earth thanks to the lack of gravity weighing them down—though I can't say if the beans tasted different. Growing food in space for longer missions, such as the journey to Mars, will be an integral part of surviving the trip since there will be more space for seeds than for cans of beans. So it's not outlandish that a botanist would be considered a mission-critical crewmember, especially for longer expeditions, where we can't fit all our groceries into the spacecraft.

Whether or not we will see boots touch down on Mars in our lifetime is still a coin toss, but that doesn't mean we can't train and prepare for that day. In fact, if you want to know what it is like on Mars without venturing into space, then take a quick trip down to Antarctica. This frigid landscape, dominated by ice, might not immediately conjure up images of the dusty red dunes of Mars, but nestled within Antarctica's snowfields exists an oasis from the mile-high ice: the McMurdo Dry Valleys.

The Dry Valleys constitute the largest area of the southernmost continent that is relatively free of ice and snow. Here, you can feel the earth beneath your boots, trek over plains of volcanic rock, and peer up at jagged treeless mountains. You won't find any cacti or shrubs or grass, nor any living animals that you could identify without a microscope. In addition to having frigid temperatures, this lonely land, one of the driest on Earth, is considered a polar desert, making it the closest earthly analog to Mars—and the perfect place for NASA to test its up-and-coming rovers.

NASA astrobiologist Dr. Chris McKay has conducted field work in this patch of Antarctica for more years than he can count, and he knows it wouldn't be an easy place to set up a farm.

> A lot of my time is spent going to places on Earth that are at the edges of the habitable zone. I'm often going to the Antarctic Dry Valleys..., the Atacama Desert, Death Valley, the top of mountains in the equator, places like that, and trying to understand life on the edge of survival. We spend a couple of months in a field station in the high mountains in Antarctica. And we get a sense for what it must be like to be an astronaut on Mars. It's not at all like in the movies. It's much more challenging and much more difficult.

Still, he is optimistic that we might one day be tending gardens on Mars, and he has even tried his hand at growing potatoes in an environment similar to that found on the red planet. He participated in an experiment in which researchers planted a spud in the dry, lifeless soils of Peru's Pampas de la Joya desert, which experiences

some of the highest levels of ultraviolet radiation on our planet. "These are the most Mars-like soils found on Earth," he says.

The soil on Mars isn't soil as we know it. When we think of soil here on Earth, we think of a medium that is full of organic material, minerals, and microbes. On Mars, however, it's more of a lifeless dust. To conduct the potatoes-growing-on-Mars experiment, the spud and the desert soils were not left in their natural environment but instead were placed inside a ten-centimeter cube that typically houses satellite hardware. The research team re-created Martian-like conditions inside the cube, accounting for temperature, pressure, oxygen, and carbon dioxide. In the end, a salt-tolerant variety of potato was the most successful at "growing on Mars." That's one small spud for Mars, one giant leap for future Martian farmers.

Farming on Mars isn't just about feeding hungry astronauts, however. Certain lichens have emerged victorious in Mars-like conditions, giving hope that they could help plants get a foothold in the environment by creating miniature ecosystems for microorganisms that are a critical component of healthy soil ecosystems. With this assistance in creating thriving soil, future residents of Mars could grow and nurture the plants and bacteria needed for certain pharmaceuticals. They could also grow fungi. Not for mushroom casseroles—no, fungi on Mars could become the very building bricks of humans' future habitats, a sort of "mycelium architecture." NASA astrobiologist Dr. Lynn Rothschild has been growing fungi inside her lab in hopes that they might one day house astronauts.

"A turtle brings its house with it but has to spend a lot of energy lugging around its shell—that's like bringing a premade habitat from Earth," she told me. "There are NASA scientists working on designing habitats you can make with [Martian] surface material, but we suggest another faster and low-equipment approach, which is using fungi to build a house." Based on her research, Rothschild believes that it would require about a month to grow a suitable structure on Mars for humans to live in. It doesn't take long after chatting with Rothschild to understand the array of possibilities for the future of agriculture on the red planet. It's entirely possible that "farm to table" might assume an entirely new meaning if that's literally where your furniture comes from.

When we eventually get to Mars, farming and gardening won't be a one-person endeavor, at least if we can avoid accidentally stranding someone there. Like going into space, farming requires a team effort. McKay believes this is something that *The Martian* pegged about the first real-life crew that will visit Mars.

> One of the things I liked about *The Martian* is that it portrayed a team working together. Even though there were problems, even though things went wrong, they worked together. In adversity they came together. That, to me, is the most important lesson that I personally learned from working in these extreme environments. And I hadn't expected it. As an engineer and a scientist, I expected that what determined our success in these environments was how good our gadgets worked and how good our gear was, and whether our tents were rated to minus twenty degrees or not. That stuff is secondary. What matters most is how well the team comes together when things go bad, and they always go bad.

Mars isn't expected to be a cakewalk for future astronauts, but through teamwork and tilling, it might just be a survivable home away from home.

8

GETTING OUT THERE

WILL SPACE ELEVATORS BECOME OUR TICKET TO THE STARS?

Chemical rockets have been our most effective method so far to launch people into space, yet it is possible to imagine that traveling offworld in the future could look different than it does now. We've iterated and improved rocket technology over the years, and we live in a world where rockets can even land themselves, but they're still the same mode of transportation that Yuri Gagarin and Alan Shepard rode into the dark yonder back in 1961. Which brings up the question, What if there are other ways to launch ourselves into space? People have developed high-altitude balloons, which have carried passengers only as far as two-fifths of the way to the edge of space. And we've tried rocket planes—aircraft that come equipped with a switch to turn on a rocket booster—which were slightly more successful.

But what if instead of floating away or rocketing up to reach the edge of space, we instead stepped into an elevator that carried us halfway to the Moon? The idea for the space elevator has been around since at least 1895, when Konstantin Tsiolkovsky, the early rocket-science pioneer who also dreamed up rotational artificial gravity, was reportedly inspired by seeing the Eiffel Tower, which had been erected a few years prior.

Well over a century later, space elevators have made their way from the pages of sci-fi novels and scientific concepts to on-screen appearances. The film *Aniara*, released in 2018, features a space elevator that ferries people fleeing climate change to a docking waypoint as they leave Earth for Mars. The film is based on a 1956 poem by Swedish Nobel laureate Harry Martinson. In 1997, the *Star Trek: Voyager* episode "Rise" featured a space elevator—the only means to safely leave a planet—in need of repair after it had been damaged by meteors. A space elevator almost made an appearance in the 2019 film *Ad Astra*, but the concept, developed by Industrial Light and Magic, was scrapped for an antenna tower that reached into space.

Most notably, a massive, towering space elevator that took generations to build plays a crucial role in the 2021 show *Foundation*, which is loosely based on Isaac Asimov's book series of the same name. Known as Star Bridge, the structure transports inhabitants from the planet's surface to an orbiting station that docks with starships. Unlike the rockets here on Earth that transport astronauts into space in an eye-opening eight minutes, the Star Bridge gently lifts travelers to orbit over the course of fourteen hours. I can only hope elevator music doesn't exist in this future.

In *Foundation*, the Star Bridge is a monument to an emperor's might and a way for many thousands of his citizens to travel throughout the galaxy without having to rely on burning massive amounts of rocket fuel. While it seems to be a much nicer, less stressful ride into space than a rocket, alternative transportation methods like this face a number of hurdles before becoming our new ticket to the stars.

According to Dr. Bradley Edwards, a physicist who received a grant in 2000 from the NASA Institute for Advanced Concepts (NIAC) to investigate these very technologies, a number of challenges are involved with bringing a space elevator to life. Despite its name, a space elevator wouldn't look or be built anything like elevators as we currently know them, mainly because materials like steel would compress at such altitudes and fall back to Earth, triggering a massive disaster. And even if you could build an elevator using lightweight materials that extended to the edge of space—around one hundred kilometers, or sixty-two miles, above Earth's surface—it would just fall to Earth due to the gravitational pull. (That would be an epic way to skydive, though.)

Since building space elevators from the Earth up isn't going to be easy, many concepts instead suggest constructing them from geostationary orbit, around thirty-six thousand kilometers away from Earth's surface, downward to the ground. In this scenario, the space elevator would be anchored at one point on Earth and secured on the opposite end with a counterweight in space, keeping the cable taut and vertical.

Edwards thinks building a space elevator halfway to the Moon, closer to two hundred thousand kilometers away, to counterweight it from Earth's gravity would be a better idea. Assuming that we could build it, what would it be like to travel on such a thing? "You'd go to an ocean platform, you'd climb into a module, and you'd feel it start moving," says Edwards. "You'd see the Earth fall away. In just a half hour or so, you'd pass up through the clouds, and you'd start to see the curve of the horizon. Another half hour to an

hour later, you'd basically be in space. You'd see stars, even in the middle of the day. Eventually, you'd be weightless. It'd be a very smooth ride; there wouldn't be any shaking."

To transport people and cargo to such great heights in a peaceful and smooth manner, a space elevator needs to overcome a couple of key problems. First is the building materials: you'd need something incredibly strong and lightweight that can also extend many tens of thousands of kilometers while maintaining structural integrity in light of the elevator's weight plus the weight of people and cargo. Second is the power system. Eager to find a possible solution, X Prize Cup and Elevator:2010, a five-year partnership between Spaceward Foundation and NASA, began holding competitions for space elevator development in the mid-2000s.

The competitions focused on the two areas in need of innovation: building a strong and lightweight tether, and developing a power system to move the elevator up and down the tether. The power system, often referred to as "power beaming," would involve the use of lasers on the ground to generate electricity for the motors, acting as a source of wireless energy transfer.

Five years of competitions netted few advancements, and NASA concluded its sponsorship of Elevator:2010 as a result. Since then, lasers have improved, and researchers have pinned their hopes on the latest developments in carbon nanotubes and single-crystal graphene. Composed of a single layer of carbon atoms, carbon nanotubes and single-crystal graphene are nanomaterials that have the capability of being a hundred times or more stronger than steel at a fraction of the weight. In 2014, employees in Google's research and development department, known as X, attempted to construct the materials to form a space elevator, but they quickly sunsetted the project after concluding that carbon nanotubes weren't yet able to be produced in a meaningful length. But today, carbon nanotubes can be created on the scale of centimeters, which sounds small but is just long enough to allow researchers to begin imagining how they could be extended to kilometers long. It turns out that constructing space elevators isn't rocket science, but it is materials science.

Yet another challenge to overcome is determining how a space elevator could dodge orbital debris for years without experiencing a significant or complete failure. Though not impossible to solve, this problem makes some experts hesitant to continue developing a space elevator for now.

The dream of space elevators may have to remain just that until a government or private entity puts meaningful funding and years of development behind it. Still, Edwards thinks it's a dream worth holding on to and is charging ahead with his work. "The space elevator will reduce the cost of getting from Earth to space. It will also allow us to take very large payloads into space very easily, very safely. Because of that, we can build cities on the Moon. We can build space stations. We can build large solar arrays in space to collect energy from the Sun and beam it down to Earth."

Without having to rely solely on rockets, the possibilities for space exploration would significantly expand. As Edwards points out, rockets, even large, modern ones, can only carry so much payload and only so often. If one day a reliable space elevator operated continuously, all kinds of human habitats in space, from recreational sites to research stations to logistical facilities, could be built at larger scales and on shorter timelines than ever before. Space travel wouldn't require being one of few individuals selected for a single excursion but instead could be a mode of transport as convenient and accessible as planes, trains, and automobiles. Until then, science fiction offers us a view into what that future might be like.

9

ASTEROIDS

WHAT'S SO INTERESTING ABOUT A BUNCH OF SPACE ROCKS?

In pop culture, asteroids typically fall into two common tropes. The first is a spaceship whizzing through a dense field of skyscraper-sized jagged rocks and lumpy boulders, narrowly avoiding deadly collisions as it outruns an enemy combatant. The second is an impending impact from a flaming asteroid that threatens all of civilization, forcing experts to assess whether we can avoid annihilation through a combination of clever technology application and creative problem-solving.

Perhaps one of the definitive asteroid chase scenes comes from *Star Wars: Episode V—The Empire Strikes Back*, where the *Millennium Falcon* dodges space rocks left and right and C-3PO informs Han Solo, "Sir, the possibility of successfully navigating an asteroid field is approximately 3,720 to 1!" (Fun fact: some of the asteroid models that Industrial Light and Magic used for that scene were actually potatoes.) Then there's the 1998 blockbuster *Armageddon*, which has been seared into the long-term memory of anyone who has seen it as the definitive, all-American asteroid-coming-to-obliterate-Earth movie. To help humanity avoid certain death, a team of oil drillers is enlisted to become astronauts so they can expertly drill into an asteroid, insert a bomb, and nuke it from orbit—ideally before it reaches Earth. From *Star Wars* to *Star Trek*, mainstream movies to indie flicks, the depiction of asteroids has been fairly consistent over the decades. So, what else is there to show when it comes to portraying big rocks in space? Are

they more than grayish, brownish, asymmetrical, cratered debris causing havoc? Are asteroids just kind of...basic?

Despite being overwhelmingly common celestial objects in our solar system, asteroids remain largely unexplored and uncharted. This may explain why their representation in cinema and television is mostly unchanged from decades ago. Asteroids come in a variety of "flavors"—made up of everything from metal to ice, olivine, cyanide, and even pieces of glass. Many asteroids aren't gray but instead are more red in hue.

"It's true that asteroids in our solar system are mostly made of rock, similar to what you would see here on Earth, but some of them are actually from the cores of failed planets," said planetary scientist Dr. Alessondra "Sondy" Springmann. "Those are made of iron and nickel, similar to the core of the Earth and Mars. Now we're sending a spacecraft to a metal asteroid with the NASA Psyche mission. It'll be the first time in history that we get to see what a metal asteroid looks like out in space."

I sat down to chat with Springmann, who specializes in researching asteroids and comets, and Dr. Michael Busch, who studies near-Earth asteroids at the SETI Institute, to dust off some of the long-standing assumptions we have about asteroids. I was curious to hear if asteroids are more complex than we are led to believe, and what we could learn from them. "Perhaps even more interesting are the asteroids around Jupiter's orbit and beyond that have frozen water in them," says Busch. "There's active debate still about where Earth's oceans came from. Did our oceans come from icy asteroids near Jupiter's orbit or the outer asteroid belt? Hopefully the spacecraft missions we're sending out in the coming years help figure that out."

While an icy asteroid colliding with this planet long ago may have turned out to be a good thing for life on Earth, today it would spell doom. The list of movies featuring impending destruction caused by a rock from space is long. And they might make you wonder what the difference is between an asteroid and a comet. Some would say that asteroids are icy dirt balls and comets are dirty ice balls. Asteroids are composed of rocky materials and/or metal but are incredibly cold from the ambient temperature of outer space (approximately -455º Fahrenheit, unless you're near a heat source); comets contain some rocky material and dust but are mostly ice. A movie may feature a comet (*Deep Impact* and *Don't Look Up*), or it may feature an asteroid (*Seeking a Friend for the End of the World*), but both are equally capable of destroying us.

Whether they center around comets or asteroids (or even a planet, as in *Melancholia*), impending-death-from-space movies usually make the characters and the audience think deeply about the troubles of our time. An early entry into this subgenre was the first sci-fi film from Italy, *La morte viene dallo spazio* (*The Day the Sky Exploded*). Released in 1958, the film imagines all the countries that possess nuclear weapons joining together to use those weapons to destroy an incoming asteroid cluster rather

than each other. Stuck in the grasp of the Cold War and the start of the Space Race, the plot offers an interesting peek into politics and global security wrapped in a thoughtful science fiction package. Similarly, about sixty years later the imminent comet impact in *Don't Look Up* conveyed a satirical commentary on the lack of action around climate change.

Although these movies are great at delivering messages about our shared planetary-scale concerns, some good news is that NASA tracks all the lurking asteroids that could be potential "planet killers," so we know it's pretty unlikely that anything like total annihilation from impact with a celestial object will happen to us anytime soon. That said, NASA doesn't track all the smaller "city killers" out there. These are objects that might not mean the end of the world as we know it but could still do significant damage. Thankfully, organizations like the B612 Foundation, cofounded by astronauts Ed Lu and Rusty Schweickart, among others, use telescopes and image data analysis to identify smaller asteroids. The goal is to work with governments around the globe to figure out how to avoid potential collisions—because it is possible to avoid a collision without sending up Bruce Willis to nuke the space rock. In fact, if we could identify an asteroid on a collision course with Earth far enough in advance, all it would take is a little nudge from a robotic spacecraft to move it out of our way.

Protecting Earth from space rocks is one thing, but what about dodging asteroids when we venture into space? If film and television are to be believed, asteroid fields show up all the time in the vicinity of people who are traveling through space. How much do we need to worry about steering around asteroids when we begin sending humans beyond the Moon?

"We don't see asteroids in fields like that in real life," Springmann explains.

> They don't tend to clump together. There are on the order of a million asteroids that we know about, and many of them hang out between Mars and Jupiter in the asteroid belt.
>
> The reason why we have an asteroid belt in our solar system is due to Jupiter's gravitational pull, not some intrinsic property of asteroids. But even in the asteroid belt, it's nothing like an asteroid field you'd see in sci-fi. If you shrunk the asteroid belt down so that the asteroids were the size of potatoes, the next closest potato to you would still be several kilometers or miles away.

So we don't have to worry about narrowly avoiding crashing into asteroids with our spaceships. However, another common trope in sci-fi turns up the fear factor to eleven: asteroids used as weapons. *Edge of Tomorrow* features formidable aliens arriving on Earth via asteroids, *Starship Troopers* features aliens using asteroids as Earth-targeted weapons, and *The Expanse* features spacefaring humans using asteroids as weapons against humans on Earth.

For now, asteroids as weapons remain firmly in the realm of sci-fi, but the notion is not entirely far-fetched. Companies and scientists are developing concepts for how asteroids could be propelled mechanically to help aid in transporting resources—*not* for weaponry—simultaneously illuminating how exciting and terrifying our future can be.

Beyond asteroids' use as weapons, *The Expanse* depicts a much broader view of our potential future relationship with them. They are portrayed in the TV series as the basis of entire economies and infrastructures, and also as shelter for human habitats. The Belters, people who live in the asteroid belt, represent a futuristic version of blue-collar workers, farmers, and miners. Although they have little connection with Earthers or the militarized peoples of Mars, Belters are a significant pillar of the collective economy since they are the ones who mine asteroids for resources and install critical infrastructure on them. *The Expanse* even features real asteroids from our solar system, most notably Eros and Ceres, which are used in the show as space stations and house small cities of people inside.

The thought of living inside an asteroid is exciting, but would asteroids as offworld stations even be possible? According to Busch, some would be easier to live on than others.

> Ceres is in the asteroid belt and is classified as a dwarf planet, and Vesta is the next largest object there. Sending humans to Ceres or Vesta is a lot more like sending humans

to the Moon. They have significantly more gravity than most of the asteroids there. The surface gravity on Vesta and Ceres is about 2.5 percent and 3 percent of Earth's gravity respectively, but it's not negligible. For example, if you jump on Ceres, you will come back to the ground. But if you jump on many other asteroids, it may take you a few weeks before you land back on the ground, if you ever do.

In *The Expanse*, we see a human habitat inside Eros that they've clearly dug out, but I'm not sure that's what you would want to do on an asteroid that small. You'd risk it becoming a rubble pile that would spin off into space, with little moons around it. You might do better to instead take parts of the asteroid off and form them into a radiation-shielding shell that you hang out inside of.

Breaking off pieces of an asteroid to form a hermit crab habitat for myself in space isn't a concept I had thought about, much less seen depicted. Springmann was quick to further explain just how complicated mining an asteroid or even creating a livable place on one would be.

> Most of the near-Earth asteroids aren't like a solid piece of rock or brick. They're actually a whole bunch of gravel and boulders loosely stuck together—what in polite company we call a gravitational aggregate

and more casually we call a rubble pile. These asteroids are microgravity environments. You can't go out to an asteroid with a shovel and just dig. The material is going to stick to your shovel like the static cling of Styrofoam. When it comes to digging into asteroids, you have to get out of thinking like an Earthling—you can't take gravity for granted.

The thought of trying to set up camp on an asteroid, or even a comet, admittedly delighted me. On deployments to Antarctica I learned how to use rocks, ice, and snow as raw materials to help secure my tent against the elements. With strong winds, below-freezing temperatures, and an icy ground to contend with, that had been challenging enough. Simply drilling through solid ice to secure a few ropes was exhausting. I could only imagine what that would be like in space, with barely any gravity and no breathable atmosphere.

The idea of asteroids as human habitats might not be as alluring as the fantasy of living on the Moon or Mars, which would be unbelievably challenging environments to make a home out of, but they are an interesting option. As we venture out further into the solar system, asteroids very well may become an easy rest stop for wayward astronauts. And given that there's so much more to explore and learn about asteroids, there's every chance we'll find something that makes us want to visit one. We just won't go jumping around on them.

10

ARTIFICIAL GRAVITY

HOW WILL WE STAY GROUNDED WHILE IN SPACE?

Artificial gravity is a hallmark of sci-fi shows set in space. The trope appears frequently on-screen due to the physical practicality of needing to film here on Earth. It's simply easier and less complex to film people performing activities on a spaceship with some form of gravity and ask viewers to suspend disbelief when it comes to wondering how that gravity was created. And given that most of us will never experience microgravity firsthand, providing artificial gravity in the setting of a story, whether it's a book or a movie, can make it easier for us to envision ourselves in tales about space travel without having to worry about all the difficulties that come with it.

Notably, *Apollo 13* is one of the few movies that has filmed actors experiencing actual weightlessness while confined to Earth. In order to achieve this, a significant portion of the film was shot within a "vomit comet," a "zero-G" plane that flies in parabolas, briefly producing weightlessness for all onboard. *MythBusters* cohost Adam Savage had a chance to film in one of these planes.

> After I was done filming for *MythBusters*, I did throw up, but vomiting did not detract from my enjoyment. I would do it again at any time at the drop of a hat. According to Tom Hanks the whole *Apollo 13* film crew and actors had a rough go as well. During those parabolas, especially when you're doing them for days and days on end, everyone throws up. No one was anything but broken from it. You'd come back wrecked from the repeated nausea, and yet the footage that came out of it and went into *Apollo 13* is unassailable and magnificent.

Watching the movie, it is amazing to think that the actors experienced true weightlessness during filming. Wouldn't it make sense for other space films to be shot in the same way? "Directors from the history of film have always worked in different ways," says Savage. "Stanley Kubrick, of course, in multiple sets on *2001: A Space Odyssey*, is using like five different hanging rigs for five different shots in sequence. So your brain is never quite figuring out the magic trick. And yet, nothing equates to that scene in *Apollo 13* when Bill Paxton, as astronaut Fred Haise, is throwing a pen across the cabin. There is no special effect that can render something that your brain just knows is real."

Trading nausea for actual shots of weightlessness works great for depicting microgravity in film, but what would it take to make artificial gravity work for real-life astronauts? Since there are no space stations or spaceships that exist today with artificial gravity, we have to turn to sci-fi to see the different possibilities for how it could be done without breaking the laws of physics.

2001: A Space Odyssey quickly became an icon in the world of cinema for ushering in a new era of possibilities when it came to special effects. The film famously showcases a massive space station, known as *Space Station V*, which is shaped like a wheel and rotates to generate a sixth of Earth's gravity, akin to being on the Moon. The concept of using rotation to generate artificial gravity has been around since long before any satellite, creature, or human from Earth traveled to space. In fact, it was written about as far back as 1903 by Konstantin Tsiolkovsky, who is considered one of the early pioneers in rocket science. The rotating wheel in *2001: A Space Odyssey* uses centripetal acceleration, subjecting the astronauts inside to the same kind of centrifugal force you'd experience on a carnival ride that swings you around and pushes you into your seat—only at a much larger scale. "This is one of my all-time favorite space films. The science in the movie holds up tremendously well," says NASA astronaut Dr. Garrett Reisman. Reisman was the technical consultant for the TV series *For All Mankind*, and he made a cameo appearance in the series finale of *Battlestar Galactica*. "And what's really amazing about it is how realistic it is considering they filmed all of this before Apollo 11, before we landed men on the Moon." He continues,

> So you might be asking yourself, why is the space station spinning, and why is it constructed as a big wheel? What you're doing is you're creating inertial forces that simulate gravity and have the same effect, because the absence of gravity causes some problems to the human body. It causes, for example, some bone loss. Your muscles, especially in your legs, start to

atrophy because you're not using your legs anymore. [That said,] if you're creating artificial gravity with a small radius, like those rides in amusement parks that make you go up against the wall, you can handle that spinning for a couple minutes. You don't want to be spending months in there, right? You need to avoid having the radius so small that you get what's called the Coriolis effect, [which] makes you kind of dizzy; you don't want that. You also need to be spinning fast enough so that if you walk in the opposite direction, you don't start floating.

The Coriolis effect occurs when items that are technically moving in a straight line appear to be moving in a curved line inside a rotating spaceship. If you were inside a giant rotating wheel and you threw a ball to a friend who was across from you on the other side of the wheel, your brain would expect the ball to travel in a relatively straight line, but instead it would appear as if the ball itself took a turn and veered off course.

Since *2001: A Space Odyssey*, other shows and films have put their own spin on the idea. The long and narrow space station in *Babylon 5* rotates around its longitudinal axis to produce the same kind of artificial gravity. *Elysium*, from 2013, featured an Earth-orbiting ringworld inspired by the Stanford torus, a 1970s NASA concept for housing tens of thousands of people. Perhaps one of the most notable on-screen depictions of the Coriolis effect comes from *The Expanse*.

Ceres, a dwarf planet that exists in our real-life solar system between Mars and Jupiter, has been converted into a space station by the digging of tunnels inside it and spinning it up to produce artificial gravity. While on the planet, a main character, Miller, goes to pour a drink into a glass, which is set slightly away from the bottle instead of directly underneath it. As the liquid pours out, its path curves, causing it to fall into the glass instead of spilling onto the floor as you might expect. Although the visual is exaggerated compared to what would actually be observed, it successfully illustrates how rotationally generated artificial gravity would be kind of weird to experience and would take some getting used to.

The Expanse demonstrates another type of physics-approved artificial gravity through linear acceleration. Long and narrow spaceships, looking like skyscrapers with rocket engines on one end, continuously accelerate at incredible speeds to produce Earth-like gravity for the crew inside. With that level of acceleration, inhabitants of Earth could probably reach Mars in a few days instead of nine months. So to ensure that they don't miss their destinations, the crew flip the spaceship partway through the journey to continuously decelerate, with their engines acting in opposition to the direction of travel, which helps to maintain artificial gravity through linear drag. When characters need to move around in sections of spaceships or stations that don't have artificial gravity, they employ magnetic boots to help them walk more normally.

While these technological concepts are all possible under the laws of physics, for the time being they are out of reach. Developing them would require extremely high sums of money and many, many years. As alluring as it might be, artificial gravity may prove to be an idea that we'll have to slowly work up to—at least until an unforeseen breakthrough emerges.

In the meantime, NASA is investing in achieving smaller versions of artificial gravity. In 2017, the NIAC program funded an investigation led by Jason Gruber into a concept that would strap an astronaut into a small-scale linear accelerator and propel them in one direction before turning them around to travel in the opposite direction. Called Turbolift, it's something like strapping yourself onto a sled and riding up and down a hallway. It wouldn't produce artificial gravity for a space station or allow you to walk around, but it could produce enough g-force that it would counteract some of the negative health effects of being in space, meaning astronauts could stay in space for longer durations with fewer consequences.

It might be a long while before we are able to take gravity with us on our way to space, but even producing a small amount could have profound impacts on how far we can go.

11

CLONING

DO WE NEED IDENTICAL HUMANS IN SPACE?

Cloning humans is a linchpin biotechnology that shows up in lots of space-related sci-fi. It's also often used as an unforeseen plot twist—*so here's a quick spoiler warning before you read on*. The examples are far ranging. The stormtroopers in *Star Wars* evolved from an army of clone troopers. In *Moon*, from 2009, a small collection of clones kept the Moon continuously staffed with working miners. In 2013's *Oblivion*, alien invaders cloned astronauts to serve as patrollers to keep the remaining humans away from their mining operations, which are stripping Earth of its resources. The television show *Foundation*, released in 2021, depicts a succession of emperor clones, each with their own unique hobbies, but all expected to rule the galaxy with an iron fist. *Star Trek: Nemesis*, from 2002, paints the possibility of using clones to deceive enemies.

Clones are almost always presented in space science fiction as a more efficient means to an end, but it is the efficiency part of the equation that snags my curiosity. Of course, any sci-fi fan will tell you that space travel and exploration are all about the need for efficiency. The less weight you have to carry with you from the surface to orbit and beyond, the better. It's resource intensive and expensive to travel offworld, so you want to keep your carry-on minimally sized. Having microscopic genetic material readily on hand to animate is the equivalent of a "just add water" freeze-dried packet of food. But there's nothing that says fertilized embryos in space need to be clones. They're the same size whether or not they're copies.

Perhaps what science fiction likes, then, isn't just the efficiency of clones, but the illusion of having control and predictable outcomes by using them. The longer you think about it, the more likely you are to wonder why clones come up so often in space sci-fi. Are they merely a way to invoke a visceral reaction toward creepy dystopian futures? Or does the science of cloning lend itself to being an efficient tool for a spacefaring civilization?

To answer my quickly multiplying questions, I turned to Hank Greely and Dr. Indre Viskontas. Greely focuses on the ethical, legal, and social implications of advances in the biosciences. As a law professor and the director of the Center for Law and the Biosciences at Stanford University, he tackles topics ranging from genetics and neuroscience to stem cell research, cloning, extinction, and assisted reproduction. Viskontas is an associate professor of psychology and a cognitive neuroscientist researching the neuroscience of creativity at the University of San Francisco.

ARIEL WALDMAN: In sci-fi, cloning can often only be done by royalty, rich corporations, or evil empires due to its complexity. In *Moon*, clones are extracted as fully grown adults from storage drawers. Similarly, *Foundation* shows adult clones floating in liquid cylinders, awaiting eventual extraction. Is human cloning actually more accessible than we think, or is it squarely in the zone of science fiction?

HANK GREELY: Some people might remember Dolly the sheep from back in 1997. She was the first mammal cloned from cells of an adult mammal. Dolly was created by taking an egg from one sheep and removing its nucleus, and then they took a frozen mammary tissue cell from another sheep that had been dead for three years. They then fused the two together, the egg without a nucleus and the mammary tissue cell that had a nucleus, and gave them electric shocks. Out of over two hundred sheep eggs, they got just *one* sheep. Scientists had been cloning frogs twenty years earlier, but Dolly caught everyone's attention, in part because people had concluded it couldn't be done. Dolly lived for six years. People got terrified. There was mass hysteria about armies of cloned warriors on the horizon, and lots of states in the US as well as [other countries] around the world passed laws against human cloning as a result.

What's interesting about clones in reality, especially within the context of how they're depicted in space operas, is that they're *almost* genetically alike, but not quite. Because scientists in real life have fused an egg's nucleus with a skin cell that has its own separate nucleus, it means a small part of the DNA in the clone—the part found outside the cell's nucleus but in little "organelles" in the cell called mitochondria—will be different from the original animal—think on the order of 16,000 different DNA bases from the skin cell compared to the

6.4 billion in the egg nucleus. But after all that work, you just get a cloned baby. Not a fully formed adult or a human with implanted memories or a set personality, just a baby.

Nobody as far as we know has done that with humans. It's illegal in most places, and it would be criminally reckless, certainly at least now because we're not really successful at it. When you clone large animals—and there actually has been a lot of cattle and horse cloning going on in the world—you get 5, 10, 15 percent disabilities and stillbirths in those animals. That's not acceptable in people.

WALDMAN: In *Moon* and *Oblivion* and *Star Wars*, we see clones that share the exact same personality. Do clones have a predictable personality?

GREELY: We actually have a way of investigating this! There are already clones among us: identical twins, who make up close to 1 percent of the global population. Identical twins are more genetically identical than clones because, unlike Dolly and cloned animals, twins have the same DNA in their mitochondria. Studies in twins show that there may be some genetic associations with things like extroversion or shyness or sports abilities, but ask anyone who has ever known identical twins if they're the exact same person, and the answer is no. They have different personalities and interests, and some have differences in things as fundamental as sexual orientation. So, they're genetically identical and absolutely not the same person.

INDRE VISKONTAS: The common sci-fi trope assumes you can grow a clone in a vat and have the clone emerge with a fully functioning human brain and artificial memories. In the movies *Moon* and *Oblivion*, we see artificial memories being used as a linchpin to keep clones in check. And this makes sense; we've evolved with remembering as a way of taking elements of our past to figure out and predict consequences in our future. But having a cloned brain ready to go out of a vat is contradictory to how the human brain works. Watch how a baby develops and grows. When they turn three or four, they start discovering that they are an entity that is separate from those around them. That's a real discovery that *has* to come with interacting with the world.

GREELY: This is one of the reasons I'm not worried very much about cloning. In film and TV, we see cloning depicted as a highly efficient method of creating identical humans. But, in reality, twins are more identical than clones and still not identical enough if you wanted a way of creating a reliably copied clone army to rule over, or as we see in *Moon*, a reliable lunar miner to do the work no one wants to do. That, and the fact that cloning still involves parents and raising chil-

dren, means that it's not really efficient at all for managing offworld outposts.

WALDMAN: What about cloning memories?

VISKONTAS: Every time we remember something, we have to *physically* reconstruct the memory, including on the microscopic level of how our neurons are activated. Even if a brain is constructed to be 100 percent identical [to another brain], if it hasn't had those experiences, it's not going to have those memories. Can you physically reconstruct that activation pattern in someone else's brain? Yes. Is it going to be the same memory? No, because that person's neurons have had a different experience interacting with the world. If you're wanting clones to have shared memories, they will need to have some shared experience of the memory.

Consciousness emerges from experience, and memories emerge from experience. Clones won't be able to see or hear or do anything if they weren't interacting with the world first.

WALDMAN: How could we realistically use cloning to benefit us when we venture offworld? Or are there alternatives to cloning we should consider?

GREELY: One scenario would be for having readily available, replaceable body parts or organs. Because if you need a new liver while you're out in space, a cloned liver is going to be welcomed by your body's immune system more than someone else's donated liver. But that's a very specific example where you don't need to clone an entire human to arrive at that solution.

Other clone scenarios we see on-screen that could be done are things like needing to jump-start a human population quickly with a capsule full of cloned embryos. However, they don't have to be *cloned* embryos. There are hundreds of thousands of frozen embryos in the US left over from IVF [in vitro fertilization] procedures, and it's a lot cheaper to use existing embryos if you're going to need to jump-start civilization—and they can be frozen for years. But what do you do in space when those embryos turn into a hundred little babies at the same time? Who's going to take care of all the babies?

In a spacefaring civilization, cloning could potentially be an advantage, but it would come with the disadvantage of restricting genetic diversity in a population. Perhaps in far future spacefaring scenarios they could add genetic diversity back in using gene editing. It sounds like science fiction, but there are groups even today that are using gene editing to try to diversify critically endangered animal species here on Earth.

What's going to really affect our future, and potentially the future of space exploration, is when we get better at manipulating stem cells. Then we'll be able to take skin cells and turn them into sperm and what are called induced pluripotent stem cells, which we could then turn into both sperm and egg cells that can be mixed together. If and when we do that, you could have kids with a single parent.

Clones in movies and television shows may indeed be a shorthand for making us reconsider the power we wield as we continue to decode our own biology. But cloning is just one of many biotech scenarios that cast cloudy skies over our uncertain future. Space is a harsh environment, and some have suggested that we may need to edit ourselves to better adapt to it. *Gattaca* demonstrates one take on this: selecting only the most genetically fit. *Alien: Covenant* arguably shows another take, where creatures and pathogens are spliced together in pursuit of producing a perfect organism. Science fiction in this way plays a vital role in making sure we pump the brakes on our journey to the stars. Venturing to distant worlds may whet our appetite for venturing into new territories in biotech, but before we do, we should make sure to have a good long look at ourselves and the implications of those explorations.

12

OFFWORLD SOCIETIES

HOW WILL SPACE CHANGE
OUR POLITICS?

Politics plays a vital role in fictional depictions of our spacefaring future—it sets the scene, builds boundaries, and explains the outlook of the characters that operate within it. It's often a necessary backdrop to quickly illustrate who's good, who's evil, and who's "just following orders." As an audience we have been primed for stories of good over evil our whole lives, but in contrast, real-life politics affects our day-to-day lives in nuanced ways that demand close examination in order to be fully understood. Paying attention to the lived experience of fictional characters can sometimes offer a more complicated depiction of space-based politics, one that challenges us to think critically about our initial judgments and continuously reassess our empathy as we journey through a show's world-building.

The Expanse became an on-screen hit quickly after it first aired, and its consistent dedication to scrutinizing the politics and social views of its characters and societies is what caught audiences' attention. The characters in the series are as complex as the multifaceted political landscape they find themselves embedded in. Storylines big and small woven throughout the plot regularly hinge on examining social struggles, speaking truth to power, and questioning identity within larger structures.

Given the success of *The Expanse*, I became curious to hear from people well versed in developing fiction about what made this show stand out and why we see so many of the political tropes we do across spacefaring sci-fi. Is the key

to creating more nuanced politics in space sci-fi just a matter of mirroring what we see on Earth? How will our politics change when humans venture into space? Do people feel less bound to society when their species is distributed across multiple worlds? I sat down with a couple of sci-fi authors who regularly think about these things.

Besides being a sci-fi author, Nick Farmer developed the Belter language for *The Expanse* TV series and the Trill and Barzan languages for *Star Trek: Discovery*. Farmer's formal education is as a linguist, and he works to support endangered and Indigenous languages. Annalee Newitz, an author of both science fiction and nonfiction, wrote *The Future of Another Timeline* and *Autonomous*, among other titles. Newitz also cohosts the podcast *Our Opinions Are Correct*, which explores the meaning of science fiction and how it's relevant to real-life science and society.

ARIEL WALDMAN: Mars sets the scene for so much of our sci-fi, but often it's a dystopian scene of militarization. We see this with *The Expanse*, *Starship Troopers*, *Total Recall*, and others. Why is that?

NICK FARMER: Fascist fantasies have always existed in our science fiction storytelling. Mars is a convenient place to put them, partially just because it's not Earth, but it's close enough. Part of the trope often includes Spartan-like training of military in difficult conditions, like we see in *Dune*—put everybody in a harsh environment that produces super soldiers. Mars is the closest thing we have nearby where we can more easily imagine only the strong surviving.

ANNALEE NEWITZ: Mars is right next door, but it's far enough away that it feels incredibly alien. And that's kind of how in the United States we've wanted to believe fascism was, like it was happening *over there*. It's true that right now we're seeing a trope of Mars being militarized in science fiction, which kind of also makes sense because Mars is the [Roman] god of war, but I think a thread of this representation of the planet comes from *The Martian Chronicles*, by Ray Bradbury, which he wrote back in the fifties, and it's his retelling of Europeans arriving in the Americas. There's already Indigenous Martians, and humans come and destroy their environment and scatter the people that are there.

So, the militarization of Mars as a sci-fi trope can tap into stories from reality about both fascism and also colonization. You even see it show up in films like *Ghosts of Mars*, where beneath the unrealistic fun and cheesiness, it really does deal with a lot of interesting social issues about colonizing a planet with a military, in this case a lesbian military who wants to build trains on Mars.

FARMER: *Starship Troopers*, for example—people think it's serious, but it's actually a parody of those fascist fantasies too.

NEWITZ: Of course, you know, capitalism and the military kind of go together, especially when you're in the early stages of occupying a space. The original *Total Recall* movie, where they're on Mars, isn't really a military dystopia; it's a capitalist dystopia. There are certain people who can't afford air, for example. Or in *The Expanse*, you see Belters who don't have access to gravity. These harsh environments are settings for producing both super soldiers and incredibly vulnerable populations.

WALDMAN: Are colonialism narratives inevitable when imagining the future of humanity in space?

NEWITZ: My question is, can you actually have a narrative about migration that isn't about colonization? The idea that you'll find or terraform a planet with an atmosphere that we can breathe and an ecosystem that we can interact with that's going to give us food…. How is it that kind of ecosystem develops with a niche for humans, but yet there's nothing in that ecosystem that is a competitor with us? Even if we just land there, we're displacing an already existing creature in that ecosystem and taking it over and maybe driving that creature out. Maybe it's not what we would consider to be awesome enough to be human equivalent. Maybe it's short and fuzzy and doesn't have language. But even if we're not directly coming in and oppressing some civilization, we're still altering the ecosystem. We're possibly throwing it out of balance. We're displacing the local life.

Of course, that's different from Darth Vader coming to a planet and just blowing it up, but there is a continuum where migration is always going to involve some kind of disturbance. That's kind of what movies like *Europa Report* highlight so well: What if we go there just to do some science, and there's sentient squids?

FARMER: A good question that gets brought up in the book *Hyperion*, by Dan Simmons, is how much we should terraform places to be suitable to us, versus how much we should adapt ourselves to best fit into other ecosystems. The discussion there kind of comes from all the things that we're thinking about in terms of our own environmentalism here on Earth. But I also think it can be extended to things like Mars, where as far as we can tell with the best current knowledge, there's no life there now. So if we wanted to colonize Mars, we would still have to make this decision. Would we try our best to change Mars? Or would we try to change ourselves?

NEWITZ: And I think we're kind of in that debate already around geoengineering on Earth. Even things like land management: How do we handle agriculture? How do we respond here in California, where the entire state is on fire? What do we do to change our practices on Earth to prevent things like that from happening? It's already part of our politics. And I think the more that we're able to manipulate our environment in a conscientious way,...we may be able to intervene in more productive ways. And it's only going to become more of an international political issue. If we decide to do one kind of geoengineering project and China decides on a different kind, can that lead to war? We're at a stage where something we do on one side of the planet can disproportionately affect the other side—especially when it comes to weather. Those kinds of politics around environmental interventions on Earth are going to extend into space and any surfaces we land on.

I think right now everybody just wants to mine the crap out of everything because they're just focused on getting as many resources as they can, be it water or precious minerals. But that's how the Europeans colonized the Americas, right? They raised money from their governments by saying, "Hey, we're going to get gold."

WALDMAN: From *Starship Troopers* to *Star Wars*, there are plenty of spacefaring civilizations living under authoritarianism. Are there fundamental realities to spacefaring societies that make it less likely for us to imagine utopias or democracies set in space?

FARMER: As they say, all politics are local politics, and I think that will only become more likely because of the realities of space travel. If you are on a generation ship [a long-duration starship that hosts generations of people who are born and die onboard], that's who you're interacting with. You may be able to send communications back to Earth, but you're certainly not getting supplies from them. You're in a microcosm that might not return to Earth at all. Depictions of empires or federations rarely address the universal speed limit of light. So when imagining our own future in space, we're often not being provoked to think about how cut off everyone will be. How are you going to actually have interstellar politics? How would that work? Even if you're traveling close to the speed of light, you may not see your representatives on the capital planet, or them you, for ten years or even many decades.

As soon as we move beyond our own solar system, however long that takes, we're not creating an interconnected society. Through most of recent human history, we've had a connection to one another. Even if it took a really long time to get the silk from China to Europe, it got there. If we actually do move beyond the solar system, we're going to have groups that are completely splintered off and may never interact with each other ever again. Connected communication is great for storytelling, but it's not really advancing our conceptions of what the future of humanity could be like.

It's additionally important to remember that science fiction is often created for our time. When you're writing about and depicting space politics, you can either approach it as a comment about what's going on now or as a thought experiment about what could happen. For most of history, it has been the former. When we're concerned about the Vietnam War, we're exploring colonization, the psychology of the soldiers, or rampant capitalism. *The Expanse* is a little unique because it is set in the relatively near future, but it doesn't necessarily have a strong ideological bent. It shows elements of fascism, capitalism, communism, and others, but doesn't stick to a single narrative about the dangers of them. That kind of allows you as a viewer to tangle with those issues, within a larger fictional setting exploring what resources are important and what society would be like.

NEWITZ: I think one of the things that's interesting about *The Expanse* is that it does try to imagine a realistic economic and political system that's really complex with multiple political factions. What I love about the Belter language that Nick wrote for the show is that it is a microdemocratic experiment in a way. The Belter language is creole, which requires pulling from all these different languages. It forces you to imagine how a bunch of different groups came together. How they had to be interacting with each other enough to form an ad hoc democratic marketplace in order to produce that language at all. That kind of detail to me is so exciting in how we depict sci-fi.

However, I feel like we're missing that in a lot of our science fiction: stories that give you a super personal tale, but also manage to use elements like these to evoke a certain backdrop at the same time. Showcasing a small story within the middle of a space opera. Using language is one way to tell these stories. Economics is another. I totally want to have a Jane Austen–like story set in the *Dune* universe. Like, what is happening in the drawing rooms of the *Dune* universe? Sure, Paul Atreides is taking drugs and learning how to ride worms….But somebody [else] is probably just trying to marry off her daughters; what's happening there? Or include a scene in *Star Wars* where Padmé is just writing white papers about tariffs on plasma exports—things that seem mundane but actually get into the meat of how these societies are put together.

How living on other worlds will ultimately change our politics remains unknown until we begin scratching farther into what some would call the final frontier. But it is not entirely unforeseen. While fiction offers us a sandbox that allows us to forecast our potential futures offworld, it is likely more prescient about the possibilities within our reality on Earth and what we might carry with us when we venture into space if we don't closely examine that reality.

13

ALIEN LANGUAGE

HOW WILL WE COMMUNICATE
WITH ALIENS?

In our hyperconnected world, it feels only natural to imagine what it would be like to try to communicate with a creature or robot from another planet. After all, we live in the same natural universe with the same stars…at least in most science fiction. Would communication, and eventually conversation, with an extraterrestrial be possible or even feasible? Some might consider the scenario impossible given that we still can't fully communicate with any of the millions of plants and animals on our own planet and are only just beginning to understand some of their languages. However, in some fictional worlds, the art of conversation is merely a question of appropriately applied technology, as illustrated by *Star Trek*'s universal translator (an electronic device), or, as depicted in *The Hitchhiker's Guide to the Galaxy*, the use of biohacking by inserting a "babel fish" directly into the brain through the ear.

Sadly, with no universal translator or babel fish on the horizon, it seems unlikely that we would be prepared to talk with a form of alien life if it came to Earth. I spoke with various experts who have some thoughts on how we might instead tackle the problem of communication with extraterrestrials. As a writer and producer of the podcast *Flash Forward*, Rose Eveleth explores how humans tangle with science and technology. Cognitive neuroscientist Dr. Teon Brooks has extensively researched how we perceive reading and language and how we're able to adapt to changes over time—from the invention of reading and writing to the advent of mass literacy across the world in the mid-1800s. I also spoke

with the psychologist and author Dr. Douglas Vakoch, who held one of the coolest job titles on the planet while at the SETI Institute: director of interstellar message composition. He is now president of METI International.

At the SETI Institute, Vakoch worked with Dr. Frank Drake, considered the father of the search for extraterrestrial intelligence. Drake came up with the idea to send phonograph recordings into space that would serve as a simplified representation of humanity to whoever might one day find them. Known as the *Voyager Golden Records*, the two albums are carried on the space probes *Voyager 1* and *Voyager 2* as they venture beyond our solar system.

ARIEL WALDMAN: So many movies imagine what it would be like to have a conversation with aliens: *Contact*, *Arrival*, *The Fifth Element*, *Men in Black*, and *Mars Attacks*. Do you think it is possible to one day have a conversation with aliens?

DOUGLAS VAKOCH: In so many movies and TV shows where the main character gets to interact with aliens in person, I get so envious. Being able to have a back-and-forth conversation would be a dream. In the search for extraterrestrial intelligence and messaging-extraterrestrial-intelligence initiatives here on Earth, we expect that this is going to be a conversation that would take decades or even centuries because any aliens we detect are going to be around distant stars.

The fact that these conversations will take time because of radio signals traveling back and forth between distant stars means they naturally lend themselves to talking about the concept of time. In messages that my colleagues and I sent out recently to a nearby star, time was one of the fundamental concepts we wanted to get across. We do that by sending pulses of different durations and then actually talking about that in the message. Being able to point to something in a conversation is key, and that's a huge challenge that we face in SETI and METI because it's so hard to point directly to anything when you don't have someone in front of you. In so much science fiction, you have the aliens right there, and you can point, and if they understand what you're doing when you're pointing, then that can really leverage the conversation.

WALDMAN: Do you think we might be asking for trouble by sending messages into space?

VAKOCH: One of the big cautions we've heard from people like Stephen Hawking is, to paraphrase, "You'd better not transmit, because whenever a more technologi-cally advanced civilization contacts a less advanced civilization here on Earth, it does not turn out well for the less advanced civilization." So there's a fear that if we transmit to the aliens and if they are more advanced than us, they could come here and annihilate us.

I think the problem with that concern is that any civilization that has the technological ability to overcome vast light years of distance to Earth could already pick up our television and radio signals. In that case, we've already blown our cover—they know we're here. If they're on their way, I think it's better to start a conversation. Show them that we make interesting conversational partners. The reality is we've been giving off evidence of life on Earth, from changes to our atmosphere, for two billion years. No paranoid aliens have come so far, so the buffer of interstellar space acts as a natural separation between civilizations. I think the concerns have been overblown, but it *is* important that we discuss the concerns. Instead of the soundbite "Don't transmit or the aliens will come get us," let's instead think about if that is a realistic concern given how much we've already told them about ourselves.

WALDMAN: Other than the astronomical time delay between sending and receiving messages, what are some of the

fundamental challenges when thinking about how to communicate with aliens?

TEON BROOKS: I'm going to focus on the movie *Arrival* because it was so compelling to have it feature linguists as front and center to the story of human interaction with aliens. While it was a story about extraterrestrials, *Arrival* showcased what linguists think about when it comes to communication here on Earth. How do we communicate language? What is universal? We as linguists always try to figure out—across all the languages that we have been able to observe on Earth—what are the things that we share?

FRANK DRAKE: I was in charge of the picture sequence on the *Voyager Golden Record*, and we were very conscious of the fact that whoever captures this thing, if they do, will be very different from us. And we must take that into account when we choose the pictures we show and the information we give. Because it could be very misleading, or it could be unclear just what we're looking at, or it might even be unclear who the intelligent creatures are on Earth. There's no statements in English or anything like that. Even using pictures, we had to assume that they see things the way we do and will interpret those images as we do. That may be wrong, but that's all we could do.

This was a very great challenge. We worked very hard to make sure there were no ambiguities, and I think we did a pretty good job. Undoubtedly it could have been more comprehensive. There are some aspects of our civilization that are not present at all. In the pictorial presentation of life on Earth, there's no dentist, there's no lawyer. That part of our lives is still a secret, and there are many things like that. And when I think about what would have happened if we hadn't had a vinyl disc in the 1970s but a CD, we could've sent the whole *Encyclopedia Britannica*.

But in a way it was a lesson to us about what's peculiar about humans versus what is fundamental to humans, when considering how to show what our civilization is like. I think the *Voyager Golden Record* does convey a great deal about us, and just the act of putting it together makes us think about what's important about us and what's special and what's good and what's bad. And so even though it may never be seen by other creatures, it's very educational for ourselves. The *Voyager Golden Records* will probably outlive all of humanity and in fact our civilization, because the Sun will swallow up the Earth around the same time as the *Voyager* spacecraft may be captured around a star system somewhere. So it becomes the only evidence that we ever existed. They're our survivors.

WALDMAN: Is written or spoken language a better choice for communicating with an alien species?

BROOKS: You see the main character in *Arrival* realize that they need to start with the basics to communicate with the aliens. They need to establish reference, establish semantics, and they have to make sure to disambiguate. If you use relational words like "here," you have to have the context that "here" means "on Earth." If you don't establish a shared context among your conversational partners, then you can't make any progress. So just to ask, "What is your mission here?" requires the creation of common ground.

And here's what is really interesting about reading and written language: it is completely artificial! Humans do not *need* a written language at all to communicate with one another—this was something that we invented. Think about the fact that mass literacy didn't really come around until about the mid-nineteenth century. It's really fascinating that we've been able to take this artificial tool, reading and writing, and map it onto something that we find to be very natural and innate to humans: speaking. How speaking and writing overlap in different ways is also interesting.

In *Arrival*, we see symbol-like writing produced by aliens using squid-like ink that

the humans aren't able to connect to the sounds the aliens are making, and there's a sense that it's a very alien way of communicating. But we also see symbolic-representation languages that don't necessarily translate to the sounds we hear. If you look at Chinese and Japanese, there is not a one-to-one correspondence between the sounds spoken and the visual aid used in the written language. Elsewhere, we know from a lot of the Latin-based languages that there's a phonetic sound-to-letter mapping, but the Latin-based languages *also* have concepts of syllables and logograms—for example, the percent symbol.

What's also fascinating about the depiction of alien writing in *Arrival* is that by being circular, they're presenting all the words at once. How do you tell where a sentence begins or ends? Well, on Earth some of our languages have case-marking—for example, ways of distinguishing subject and object and possessive words. We often talk about the linear order of words: subject, verb, object in some languages. But if you have languages with case-marking you have the liberty of having the order be entirely different because you'll know how the words fit into an overall context. So if aliens use case-marking like we do here on Earth, it won't matter exactly which part of that written circular diagram you start with, because it could still be translated by using those concepts.

WALDMAN: Alien communication as depicted on-screen has taken many different forms over the years. Notably, *Close Encounters of the Third Kind*, while famous for using five tones of sound as a type of communication, portrayed alien telepathy with humans. Would alien telepathic communication be entirely in the realm of science fiction?

ROSE EVELETH: Telepathy has a fascinating history. It is an idea that we've wanted to believe in for so long. The word "telepathy" dates back to the early 1900s and to this guy named Frederic W. H. Myers. It was during a time when a lot of scientists were grappling with the fact that science was starting to explain things in nature that they had believed were "divine." How do you both believe in science and also believe in God? If they could prove telepathy, then in a sense they could prove the existence of God, at least partially, and solve this dilemma.

Inspired by the emerging world of invisible information at the time, like X-rays and sounds transmitted through telephones, researchers would ask, "Well, why wouldn't the human body emit some kind of signal that we can't see?" And they would write things like, "The brain probably generates electrical signals, and if so, then we could read them." Which is

exciting because, yes, it does! But not at the strength you would need in order to read them out of thin air. So around this time lots of experiments were done to try to prove telepathy. The field of telepathy was actually hugely instrumental to science for a time, and it basically developed the double-blind study. Very famous people like Thomas Edison, Alexander Graham Bell, Nikola Tesla—they all wrote as if in just a couple of years they were going to prove telepathy.

But today we don't call it telepathy; we call it brain-to-brain communication, because that sounds fancier. Which is funny because that's the same reason why Frederic W. H. Myers invented the word "telepathy" in the first place, saying "mind reading" was considered too "woo-woo." Today there are all these interesting experiments, like one where they try to connect two people via their brains...and see if they can play a game together. There's an experiment where people try to drive a car using a brain-connected headset to control the car. Some experiments have people watch movies while the researchers look at their brain activity and try to reconstruct what they're watching.

What's interesting from a linguistics perspective is that brain-to-brain communication is never trying to get people to transmit a word. They're trying to get people to transmit images or even just frequencies.

Words are actually really hard and complicated to figure out how they work in the brain. So, there's lots of really interesting modern research out there because it *is* such an alluring idea that you could beam a mental picture over to someone. It would just make everything so much easier.

Contact's depiction of a woman whose career was to find and decode alien messages inspired and catalyzed countless viewers to pursue careers in space exploration. After I watched the movie, what stuck with me was the mind-blowing realization that aliens might not even have physical forms when we encounter and talk to them. I've always loved imagining how we might communicate with intelligent aliens, likely because I find communication to be an infinitely creative endeavor that can draw inspiration from so many aspects of our lived experience and history on Earth, in terms of our interactions both with each other and with nature. In thinking deeply about how we might interact through sounds, pictures, words, and even brain-to-brain exchange, we learn more about the aspects of our communication with one another that we take for granted, and in turn more about how weird *we* might just be in this vast universe.

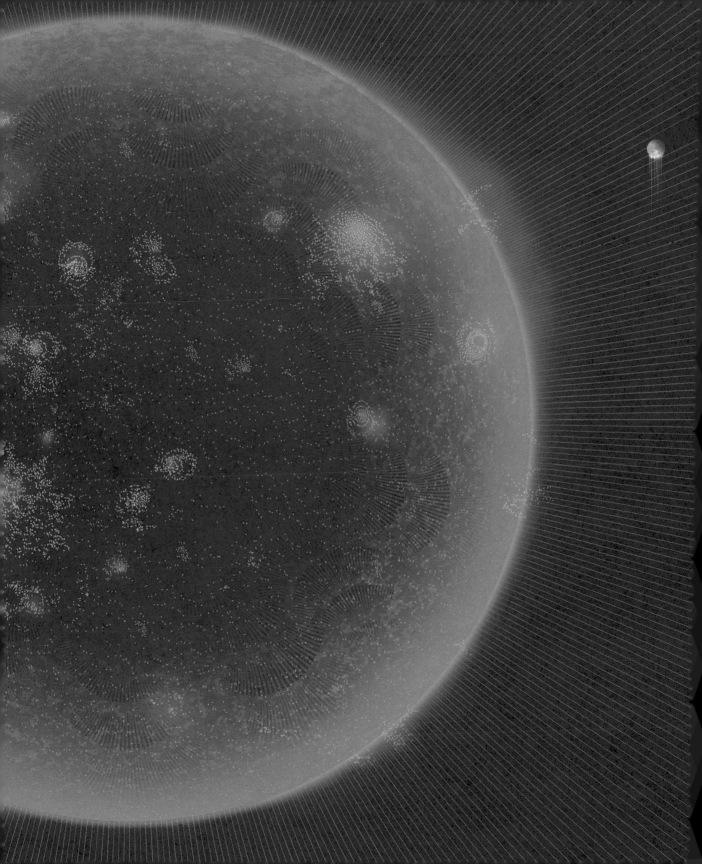

14

THE END OF EARTH

COULD WE SURVIVE THE DEATH OF OUR SUN?

Whether we like it or not, several hundreds of millions of years from now, Earth will be a dry husk of its former self. Our star will die, expanding on its way out, and much later will eventually engulf the whole of Earth. This inevitable, looming destruction invites reflection on our place in both space and time, but it also can be a source of inspiration in figuring out if humanity could become a multiplanet species. An extra planet to live on could give us an extra lottery ticket to ensure our long-term survival.

Plenty of movies and TV series take us on journeys that expand our concept of living on planets other than Earth, like *The Expanse*, *Star Wars*, and *Dune*. However, few tackle the reality of living on a planet with a sun that will eventually die, threatening humans' existence. That's where *The Wandering Earth* differs from the pack. Released in 2019, this Chinese sci-fi film turns the common trope of humans becoming a multiplanet species on its head by asking, What if we moved the Earth itself to a different star instead of humans leaving the planet behind? Of course, the most suitable star the characters identify is 4.2 light years away, and the only way to get there is by using thousands of engines built and erected across half the Earth to propel the planet out of harm's way. Oh, and they have to make sure they don't collide with Jupiter, because they decide to use the massive gas giant as a gravitational slingshot to further accelerate Earth out of the solar system. No big deal, right?

Inspired by the novel from award-winning author Liu Cixin, the film was a hit and became China's fourth-highest-grossing film of all time—and no wonder; it's a thrilling action-adventure movie. In addition to exploring what would be needed to physically move the Earth, the film also looks at how the technology and the act of moving the planet would drastically change the very nature of our home. Hint: it gets very, very cold. It offers an incredibly distinctive and interesting look at a problem that many science fiction fans might not have even considered.

Another approach to dealing with a dying Sun came in the film *Sunshine*, released in 2007. Instead of portraying people trying to help Earth escape the Sun, the movie depicts astronauts being sent on a mission to restart a rapidly *fading* Sun that is threatening to freeze out the Earth. Equipped with a bomb the size of Manhattan, the astronauts are propelled into space with the hope of creating a strong-enough reaction to jump-start the dying star. The movie is a visual feast of solar imagery, from the deep oranges and yellows that wash certain scenes with an almost heavenly glow to the unforgettable moment of seeing the astronauts silhouetted against the grand and mighty Sun. If you are looking for a beautiful sci-fi flick, this one delivers.

Both *Sunshine* and *The Wandering Earth* approach these fictionalized conundrums with the Sun in whimsical, creative ways. However, unlike in the films, our Sun will not dim out or engulf the Earth within a mere hundred-year timespan;

Sunshine is set in 2057 and *The Wandering Earth* in 2061. The eventual death of the Sun is a future event that, although not at the forefront of our minds as we conduct our everyday lives, silently contextualizes our lived reality. Even if we overcome all the odds and manage to continue progressing as a species for millennia, this one huge problem will still loom on the far horizon, and it will force us to either adapt or surrender.

With all this in mind, I spoke to two stellar astrophysicists from the University of California at Berkeley to probe them about what's really possible when it comes to surviving the death of our star. No strangers to cosmic explosions, Sarafina Nance studies supernovas, and Dr. Gibor Basri is credited with discovering an entirely new type of celestial object known as a brown dwarf or "failed star." Basri also occasionally teaches classes on the science of science fiction, so he's well versed in the wonderful interplay between the two.

ARIEL WALDMAN: People often cite the eventual expansion and death of our Sun as one of the main reasons we need to become a multiplanet species. Help break the problem down for us. What is going on in the Sun that makes it such a danger to Earth in the future?

GIBOR BASRI: It's all made of hydrogen and it's extremely massive—so much so that its own gravity crushes its core to conditions that generate nuclear fusion. It's fusing hydrogen to helium; the helium then settles to the middle of the Sun, where it is not able to fuse further for the time being. And so the core of the Sun is slowly turning from hydrogen into helium, and that makes the hydrogen outside that core burn even hotter. As a result, our Sun is slowly getting brighter, and in half a billion years or so, it's going to be bright enough to boil the Earth's oceans. Ten billion years from now, it gets a whole lot brighter, and it'll turn the Earth into a lava planet.

SARAFINA NANCE: Between half a billion years and ten billion years [from now], the Sun is definitely going to expand. But it has another five billion years to be on the "main sequence." Meaning it's just happily chillin', fusing that hydrogen to helium. And then as it reaches the end of the main sequence, it's going to expand significantly. Large enough to just about engulf the Earth, but not much more than that. So we don't have to worry about it gobbling up our entire solar system, and we don't have to worry about it engulfing the Earth anytime soon.

WALDMAN: The movie *Sunshine* explores the fictional scenario of our Sun rapidly dimming and the efforts of humanity to

97

solve the crisis. If we discovered that our Sun was fizzling out like an old light bulb, would we be able to do anything to rekindle its flame?

BASRI: Unfortunately, not a chance. Even with a bomb the size of Manhattan, as we see in *Sunshine*, and even if it was a bomb made of antimatter, it wouldn't be enough. The Sun is a bomb that's phenomenally more powerful than any bombs we can come up with. The luminosity of the Sun in just a single second is enough energy to power our current civilization for fifty thousand years. So it's a whole lot of energy. And the Sun actually has really powerful natural explosions on its surface quite regularly, which doesn't bother or disturb it at all. We could certainly blow off a whole bunch of gas on the side of the Sun that we detonate a bomb on, but that'd be about it—and the Sun does this by itself regularly; they are called coronal mass ejections.

All the action in the Sun is inside the core. People often recite the known fact that the light we see coming from the Sun takes eight minutes to get to Earth, but that's not actually the important part. The important part is that same light was produced by nuclear fusion about a million years ago. It takes around a million years for the gamma rays that are generated when that nuclear fusion takes place to physically work their way out from the core of the Sun to the surface. Any light we see coming from the Sun now is in fact ancient history. If the Sun is dimming out, that means that a million years ago the core had a problem.

WALDMAN: In *The Wandering Earth*, we see humanity attempting to move the Earth to a different star. As fantastical as that sounds, could we do it?

NANCE: There'd certainly be some monumental challenges to overcome. *The Wandering Earth* imagines a scenario in which we place thousands of rocket engines, each taller than Mount Everest, on one side of the Earth to transport it to another star. One of the problems is just getting the fuel for those engines. You'd lose something like 95 percent of the mass of the Earth just to fuel all those engines. Another problem would be anchoring those engines since they'd most likely end up falling back into the Earth as they pushed against it. But even if we were successful in engineering solutions to those problems, the atmosphere of Earth would likely be ripped off as it propelled through space. Surprisingly, though, the magnetic field might not be affected since the convection in Earth's core could theoretically still work while it was being moved.

WALDMAN: *The Wandering Earth* certainly demonstrates all the many things that would go wrong by trying to move the Earth in this manner. Are any other options on the table?

NANCE: If we did want to physically move the Earth, there are other mechanisms to consider as alternatives to just strapping rocket engines to it. We could try using a light sail, which uses the light from the Sun as a propulsive force, or using the ballistic properties of meteors to nudge us along. But those efforts would be considerably smaller and take on the scale of millions of years to achieve the kind of movement seen in the film.

The idea of the death of our Sun and the eventual extinguishing of the only home we've ever known might strike fear into the hearts of the existentially minded, but our cosmic fate can clearly also inspire creative possibilities that explore how we can tackle insurmountable challenges. None of us will experience the death of the Sun in our lifetime; it's unlikely any human will. But we are all experiencing the creeping crisis of climate change, and the many people that come after us will as well. Films like these offer us a way to evaluate the pros and cons of taking extreme, unorthodox measures to solve planet-threatening problems, and the sacrifices that are or aren't worth making for the sake of preserving humanity. In contemplating our star's last gasps, we are forced to confront not only our own mortality but the mortality of all humanity, and whether we can or should do anything about it. Personally, I'll also take it as a good reason to appreciate living on a planet whose star will support life for long enough to figure out our next move.

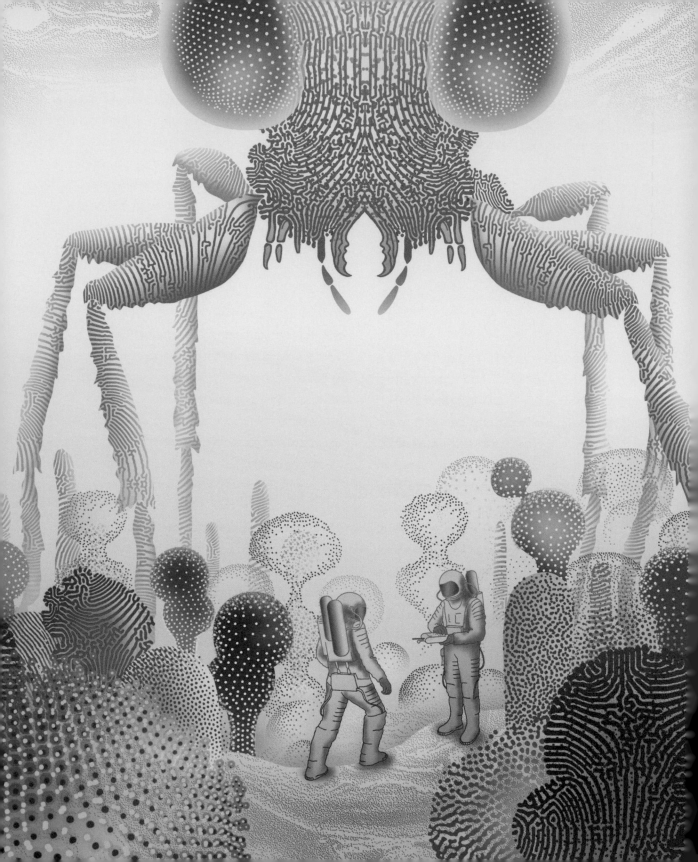

15

GIANT ALIENS

COULD THERE BE COLOSSAL KILLER INSECTS LIVING ON OTHER PLANETS?

Aliens on-screen have come in all kinds of forms: beings of pure energy, furry friends, robots, oozing animals, giant amoebas. Even though the trope of little green men with pear-shaped heads and large eyes has persisted for generations, envisioning what aliens might be like is an enduring area of imagination and creativity for both science and science fiction. In science, the field of astrobiology consumes itself with studying life "at the edges" here on Earth, allowing us a broader idea of what could exist out there, on planets and moons, both inside and outside our solar system. Increasingly, science fiction looks to astrobiologists and biologists for an informed take on how life as we know it—and sometimes as we *don't* know it—might exist.

In 1997 *Starship Troopers* took a stab at depicting a world of aliens that felt familiar enough, except for one aspect: the creatures' size. Inspired by earthly insects, the film portrayed gargantuan bugs capable of destroying Earth from far away. But this wasn't when huge insects made their screen debut. In 1902 the short film *A Trip to the Moon* featured Selenites, insectoid inhabitants of the Moon. *Honey, I Shrunk the Kids*, from 1989, featured gigantic bugs that came to life through the magic of mechanical puppeteering and stop-motion.

To create their extraterrestrials, the makers of *Starship Troopers* called on the animation studio that had produced *Jurassic Park* because of its success in revolutionizing the combination of computer graphics with animatronic puppets. Still, depicting tons of alien insects proved challenging and required too much computing power, leading to a few cut corners, notably the killer arachnids, which only have four legs instead of eight.

To sort fact from science fiction when it comes to the insect world, I spoke to two entomologists, scientists who study insects. Dr. Lauren Esposito is arachnology curator at the California Academy of Sciences, where she studies spiders, scorpions, and their venom. Dr. Brian Fisher studies African and Malagasy ants as the academy's curator of entomology. To say these two researchers love insects is an understatement. After watching *Starship Troopers* they both confessed to having a hard time with the "bug-stomping" scenes.

ARIEL WALDMAN: *Starship Troopers* **is known for imagining colossal killer insects on another planet, but is it even possible that insects, alien or not, could be larger than an elephant?**

LAUREN ESPOSITO: I can tell you that the ancestors of modern-day arachnids on Earth were huge! There used to exist creatures known as sea scorpions, and these things were massive. They were, like, two meters long. Certainly not quite as big as the arachnids seen in *Starship Troopers*, but still sizable. They were really like the grizzly bears of the ocean, and they were predators. It was these sea scorpions that first started coming onto land and became amphibious. It's been hypothesized that they were going into rivers and eating salmon out of rivers, just like grizzlies do today.

BRIAN FISHER: Having massive insects inhabiting an alien ocean would make a lot of sense. One limitation of an exoskeleton is that it's pretty heavy, but if you're in the ocean you've got some buoyancy. Maybe on the planets we see in *Starship Troopers*, the gravitational pull is less, so their bodies could support a larger exoskeleton. On Earth, why and when in our planet's history we've had large insects has been in part due to oxygen levels. Insects passively respire, meaning they don't have a lung, so they need to absorb enough oxygen to keep their muscles working. If you increase the percentage of oxygen in the air, you can absorb it quicker and be a larger insect.

In the past, Earth had ants that were as big as rabbits thanks to periods with more oxygen. Insects' "body plan," so to speak, is pretty efficient since it requires low input [oxygen] and doesn't waste that energy on heat and requires less water. Because insect skeletons are on the *outside* and the muscles are on the *inside*, that provides a Newtonian advantage. You get great leverage, and you can lift more

for your size. In many ways it's superior to our human body plan of putting muscles on the outside, bones on the inside, and then just skin to hold it all in. So it's not entirely made up that insects could actually get this large. It's conceivable that somewhere in the universe there's another planet that is similar enough to Earth in that it contains oxygen and has similar kinds of habitats, like oceans and land and things that are primary producers of oxygen, the equivalent of our plants.

WALDMAN: When we look out into the galaxy, we often see planets with environments that are much more extreme than here on Earth. Would massive alien insects be better equipped with the survival skills and hardiness to thrive on an exoplanet?

ESPOSITO: Just look at scorpions, which are a kind of arachnid. There are scorpions in caves one hundred meters below sea level, and there are scorpions in the Alps. There are scorpions in every ecosystem on Earth...with the exception of the Arctic and Antarctic. So here's just this *one* organism that has persisted for 450 million years on this planet that's really good at adapting to incredibly harsh environments: from the driest desert to the most humid tropical forests, all the way up to freezing temperatures, where there's permafrost almost year round.

FISHER: Some insects are able to survive in the tropics and then, in a sense, "shut down" during the coldest periods thanks in part to having their skeleton on the outside and being able to temporarily close off all the places they breathe through. But other [insect species] can naturally produce and inject antifreeze into their whole body in order to survive. So, there's all these really ingenious adaptations that we see in insects and arachnids just here on this planet that would allow them to withstand some incredibly harsh conditions like I would imagine on other planets.

Another survival skill we find in insects, and the brilliance of ants especially, is their ability to work together. People say they're so great because they work together like a social family, but it's more than that. There's a cooperative intelligence and a division-of-labor intelligence that's coded into their significantly different sizes. You can have large and small ants that are almost a hundred times different in size, and those differences allow them to take on different jobs, just as you see in *Starship Troopers*. What's even more fascinating is if you took an ant colony and counted all the neurons in it, it'd be about the same number of neurons a human has!

WALDMAN: But these alien insects in *Starship Troopers* are doing much more than surviving—they're equipped to the nines with all sorts of bioweaponry that puts human-made artillery to shame and makes for some pretty gory deaths. Is there anything that terrifying on our own planet?

ESPOSITO: There are definitely multiple different kinds of bugs that were depicted, and in each of them you could see where the [filmmakers] drew their inspiration from in nature. Some of them had kind of beetle heads, and others had sort of a dragonfly or a crustacean type of body, and some of them look like weevils.

FISHER: In the movie, we see things like a fire-breathing beetle, which was clearly a bombardier beetle. In nature, there's no fire, but instead it uses two chemicals inside its body that it sprays together, and when the chemicals mix, they blow up, somewhat like a fireball. That's not fiction; that's real! As a defense the beetles throw two chemicals together, and those explode.

ESPOSITO: And they push the chemicals out of the air holes in the side of their body. They have little glands right next to their air holes, and they just push the chemicals out, and the chemicals spray into the air when they're getting attacked. And it's like a chemical burn; it's acid. We also see in the movie this beak mandible thing. When I saw those, I thought, Hercules beetles! They have horns that they use for male-to-male combat. They use them in sparring over access to females, which don't have horns.

FISHER: In nature, there's a kamikaze ant that blows itself up when it's being attacked to take down its predators. But in the film we also see piercing, sucking mouthparts, which are the usual tool of some insects. When they went into the human's brain, that was classic Reduviidae, an ambush predator known as the "kissing bug" or the "assassin bug." It's a bug that has piercing, sucking mouthparts that go into its prey. It's terrifying.

What I enjoy so much about space sci-fi is its ability to make us think about how alien-like our own planet is. It's difficult for us to dream up what alien creatures might be like without taking a page from nature on Earth. Far from a limitation, that's a golden opportunity to look at our home with fresh eyes and see aspects of our world that often go overlooked. Battles between Earth insects may occasionally be displayed on nature documentaries, but it is important to remember that they take place all around us each day. Because the bugs in *Starship Troopers* are inspired by insects' real-life weaponry, the film, perhaps unintentionally, is similar to *Honey, I Shrunk the Kids* in that it asks us to appreciate the natural ingenuity of this planet's creatures. If we do someday discover colossal killer insects living on other planets, maybe we'll realize that they're not so alien after all.

16

BLACK HOLES

HOW DO WE DISTINGUISH SCIENCE FROM FICTION?

Black holes draw us in.

Yes, that may be because they are the most massive objects in the universe, with a gravitational pull so intense that even light itself is unable to escape. But we also gravitate toward them because of their enduring scientific allure. They are one of the most fascinating objects in the natural universe, but they somehow also feel *supernatural* to us. Even as science continues to make groundbreaking discoveries about black holes, they will remain partially cloaked in mystery.

So it's no surprise that black holes are a massive magnet for science fiction plot devices and scenescapes. While many stories have used their mystery as a kitchen sink for far-out ideas, the actual science behind black holes may be just the thing to inspire an entirely new wave of imaginative sci-fi.

From the big screen to the small screen, black holes have been used to hide cities, spark psychological experiences, destroy entire planets, and connect characters to different places in space-time. Black-hole plots abound in *Star Trek* (notably the 2009 reboot), *Event Horizon*, *Andromeda*, *Stargate SG-1*, *Foundation*, and many other movies and television series. In its 2021 season, *Star Trek: Discovery* prominently featured gravitational lensing, a common visual distortion seen in telescopic images when a black hole warps how light reaches the telescope. Place a typical wine glass on top of some text and you'll see a similar distortion through the base of the glass. The main plot of the episode

centered on what was suspected to be an erratically moving, hard-to-locate black hole that was gobbling up planets in its path.

Disney's *The Black Hole*, a *Star Wars*–esque space opera from 1979, attempted to capitalize on the growing scientific interest in black holes, which had only been named as such a handful of years earlier, in 1967. Being one of the last films to rely heavily on manual effects and one of the first to use computer graphics, it straddled the technological beginning and end of two film eras. The title sequence depicts a black hole, created using programmed vector graphics, to show a funnel-like warping of space-time. In its day it was the longest computer graphic sequence to appear in a film. Later in the movie a black hole is produced by filming a whirlpool of paint inside a Plexiglas water tank.

Thirty-five years later, *Interstellar* famously enlisted theoretical physicist Kip Thorne to help ensure that the film presented the most accurate depiction of a black hole to date. Instead of a circle, the movie depicts a sphere with intensely glowing light orbiting around its equator, what Thorne calls the "shell of fire." These are photons that have been trapped in the black hole's orbit but haven't gone past the event horizon where light can't escape. The equatorial "hula hoop" of light is warped to appear as though it wraps around the top and bottom of the black hole, representing the gravitational-lensing distortion of the light as we would see it.

The science fiction world took notice. Since then nearly all portrayals of black holes have looked eerily similar. However, science continues blazing a trail forward, and following *Interstellar*'s debut the scientific community has made giant leaps in both understanding and visualizing black-hole phenomena. In fact, a year after the movie's release, Thorne and his colleagues directly detected gravitational waves from a black-hole collision for the first time, winning a Nobel Prize. Shortly thereafter, hundreds of astronomers collaborated to use nine telescopes, collectively known as the Event Horizon Telescope, to capture the very first photograph of a black hole—or, to be more precise, a black hole's shadow and its magnetic fields. The image appears as a swirling inferno punctuated by a pitch-dark circle.

Despite the on-screen popularity of black holes, few know what a black hole actually is. How much do scientists understand about the secret lives of these space-time singularities? Astrophysicist Dr. Fatima Abdurrahman is the person to ask. Having studied black holes for years, she has also worked on improving the instruments used to detect them, and she has taught a course on detangling the differences between black holes in sci-fi and reality, so she knows a thing or two about science fiction.

ARIEL WALDMAN: Many of us are taught that black holes can occur when an extremely massive star dies and collapses under its own weight. How would you describe them?

FATIMA ABDURRAHMAN: What I find to be an important aspect of understanding black holes is to first have a backup definition that helps you think about them. For black holes, it's the concept of escape velocity. The reason a rocket ship can leave the Earth's gravitational pull, but I can't leave it if I'm jumping up and down, is that the rocket ship has a velocity high enough to surpass a threshold set by the Earth's gravity: an escape velocity. So, anything traveling at speeds *above* this velocity can leave the Earth's gravitational field.

That velocity is determined by how much mass an object like a planet or star has, and how big it is. So a black hole is a dense, massive object for which the escape velocity *surpasses* the speed of light. Since the speed of light is the universal speed limit on everything—nothing can travel faster than light in a vacuum—the escape velocity of a black hole means that nothing can escape it once that item enters the event horizon, which is essentially the imaginary shell surrounding the black hole where the escape velocity is the speed of light.

WALDMAN: Black holes always seem to invite intrigue, which in turn inspires these mysterious narratives about them on-screen. Are they still mysterious to science?

ABDURRAHMAN: It's true that there's still a lot of mystery about black holes. The challenge with studying them and trying to say anything definitive about them, particularly internal to the event horizon, is that the vast majority of astronomical information is gathered through looking at light, with exceptions for neutrinos and gravitational waves. How do you really understand something without light? We often infer that something *can* exist and have certain characteristics based on scientific theories, but then we still don't *know* for certain, so black holes become a fun thing to speculate about.

WALDMAN: In Disney's *The Black Hole*, and in countless other films, there often is a spaceship hanging out near a black hole, usually using an antigravity field or some other technology, or just staying outside the event horizon to keep from falling into its grip. Antigravity fields aside, is it possible to escape from a black hole?

ABDURRAHMAN: In the movie, we see the characters aboard the spaceship constantly pushing against the force from the black hole. So that means we've established the idea of escape velocity. Right at the event horizon the escape velocity would be the speed of light. A little bit farther away from the event horizon, it would still be very high—say, 90 percent of the speed of light or on that order of magnitude, depending on how far away they are from the black hole.

If you're able to travel faster than the speed of the escape velocity where you are, there's nothing to say you couldn't outrun a black hole. So it's science fiction because we haven't developed an ambitious technological feat that propels us close to the speed of light. But it's realistic to say if that technology existed, you could resist the pull of a black hole—though other things being pulled in by the black hole might collide with you before you'd even have to worry about crossing into the event horizon.

WALDMAN: *Interstellar* represented a huge leap forward in how black holes are depicted in science fiction. It's safe to say that it inspired and impressed both scientists and filmmakers. So do you think we are depicting black holes accurately now?

ABDURRAHMAN: In the past, we've seen black holes depicted as discs, but depicting them as spheres is more accurate. Black holes are radially or spherically symmetric objects for the same reasons that most things in space are spheres—planets and stars, they're all spheres. It's because gravity is symmetric, and thus how things collapse gives you lots of spheres.

One of the things that really struck me about *Interstellar* was the complexity and variety of the visual depictions of the black hole. They wanted to think about how a four-dimensional object like a wormhole would appear in a three-dimensional space, so they made the wormhole that transports the characters to the black hole appear as a sphere rather than a two-dimensional circle. They then not only depicted in great detail what a black hole would look like if you were next to it, but also what it would look like if you were peering at it through a wormhole. They meticulously calculated these things using Einstein's equations of relativity and a number of theories that have been developed since then.

WALDMAN: Black holes are often presented as scary forces of nature that can massively distort time and destroy everything in their path—is that true?

ABDURRAHMAN: Plot-wise, the idea that black holes just go around gobbling up everything requires some context. One of the little facts I always pull out that surprises people is that if our Sun right now turned into a black hole of the same mass, we would keep going around it the way we do now. We wouldn't get sucked in. We would continue orbiting this newly created black hole because, even though black holes have these extreme gravitational properties when you get close to them, at a distance they still behave like any other object with mass. So black holes are not as aggressively violent as people tend to think.

Separately, in *Interstellar*, we see black holes used as a plot device for the passage of time being different in different locations— that effect is called time dilation. Basically, there's an effect on sensory relativity that says when you're in very, very strong gravitational fields, the way time flows can be distorted, so if you are close to a black hole, the passage of time for you goes more slowly. In the movie we see Matthew McConaughey's character travel back to Earth, and everybody is much older since time slowed down for him while he was near the black hole. It seems pretty outlandish, but that's actually a very rigorously tested, theoretically supported idea that we classify as true!

WALDMAN: Do you think there is more room for black holes to evolve on-screen?

ABDURRAHMAN: There's so much more we're discovering that we could be depicting. Gravitational waves, for example, are ripples through space and time that we often detect when two black holes collide and create large ripples we can measure here on Earth. This completely groundbreaking, revolutionary discovery in physics is fundamentally changing the way we study black holes. Now the electromagnetic spectrum is no longer the only way we can glean information from events in space, and that opens the door to expanding how we depict black holes.

I feel like this is such a ripe area for sci-fi in the future because it's a whole thing that we've not really considered before. Think of how much of modern technology is about being able to manipulate electromagnetic fields. What if we could manipulate gravitational fields? It's a fun thought experiment.

In 2015 we first directly detected gravitational waves that were generated by a collision of two black holes. To be alive on this planet during that discovery is nothing short of thrilling. A century passed between Einstein predicting the existence of gravitational waves and our detecting them here on Earth. In the wake of the discovery, a new epoch of astronomy has begun, one that is no longer entirely dependent on the spectrum of light—from gamma rays to visible light to radio waves—to discern objects in the cosmos. Instead, we can now measure when a gravitational wave from a cosmic event rolls past us, stretching our planet by a fraction of a proton. It's also exciting to live during a time when filmmaking, computation, and graphics are ready to take on the challenge of portraying the latest science on-screen. The combined effect may result in a new wave of researchers and explorers who will inspire sci-fi with their work for decades to come.

17

SUSPENDED ANIMATION

CAN WE SLEEP OUR WAY THROUGH LONG SPACE JOURNEYS?

It is known by many names: hypersleep, cryogenic sleep, hibernation, and suspended animation. And there is no doubt that the concept is alluring: to survive long, relatively boring journeys in space, we cozy up in a little pod the size of a first-class airline sleeper seat as gas or liquids billow in to slow down our metabolism and make us fall asleep. Aboard the space shuttle and the International Space Station, approximately 78 percent of astronauts take sleeping pills most nights to help them get their daily eight hours (rather than to pass the time).

In *2001: A Space Odyssey*, hibernation pods provide life support to astronauts on the way to Jupiter while the spaceship's sentient artificial intelligence, HAL 9000, and other astronauts monitor—and, crucially, *control*—their health. In the films *Passengers* and *Pandorum*, hibernation pods are used to keep thousands of crew and passengers in a state of suspended animation to stop them from aging and allow them to survive a 120-year-long journey to a new planet. The pods show up in *Alien*, too, for similarly long journeys. Of course, as wonderful as the pods sound, nearly all the versions that appear in sci-fi films and TV shows have one thing in common: something goes horribly, irrevocably wrong. Computers kill people. People wake up too early to survive the journey. People suffer memory-loss psychosis. These are a few of the horrific situations that can apparently happen when humans enter hypersleep. Still, the idea of sleeping our way to the stars remains alluring enough for research-

ers to continue investigating how we could actually pull it off.

In 2016's *Passengers*, starship inhabitants enter individual hibernation pods that provide minimally invasive life support. There are no tubes, no freezing, no preservation in liquid—as are often seen in other sci-fi depictions—just some regenerative, antiaging drug therapy and a nice place to sleep, more or less. Providing technical analysis and answering questions from the press and public during the film's release was aerospace engineer Dr. John Bradford, who runs SpaceWorks, an aerospace design contractor for NASA and the US Department of Defense.

SpaceWorks has been awarded multiple grants from the NASA Innovative Advanced Concepts (NIAC) program to study the viability of human hibernation in space from an engineering point of view. What better way to ensure that your movie's fictional hibernation pods are up to science snuff than by having an expert weigh in? Unlike in the film, Bradford's real-life version of hibernation pods would lower the person's body temperature through ambient air and drug therapy, or through a nose tube pumping in dehumidified air plus an evaporative coolant, combined with sedation to induce hypothermia. In fact, this technique is currently used for surgery patients who have experienced traumatic injuries. By cooling their bodies down through what's known as therapeutic hypothermia, surgeons reduce their need for oxygen and thus can buy some time before brain damage or heart failure sets in.

Bradford says that *Passengers* "erred on the nonfreezing sort of approach [in its depiction of hypersleep], which is what we think is the more viable option." He points out that there are companies that freeze people for other purposes, but there are "no real plans and no ideas for how to bring people back from that. We think that's a much harder problem." And that isn't the only obstacle. If you don't stop all cellular activity through freezing, he says, there would still be some aging, but in hypersleep, with advances in medicine and in reducing metabolism, the aging would happen much more slowly than normal. It turns out that if your body's temperature is lowered by about 10 degrees Fahrenheit, your metabolism will be reduced by around 70 percent, perhaps granting you extra time on a space journey, but by no means causing you to arrive without having aged at all.

So there are a few kinks to consider when getting into the pod, but what about problems and disturbances that happen outside of it? In *Passengers*, unexpected turbulence caused by an asteroid collision causes a breakdown in the starship and its handling of life support, waking some of the passengers. "You have to have contingency plans for a malfunction or if a micrometeorite or debris hits the ship, for example," said Bradford. "How quickly can we warm people up? The warming process is slower than the cooling-induction process. If everyone wakes up, they're all in a pretty small habitat. It's only six months, but they'd be in for a rough ride. That's a scenario we have to consider."

Even the acts of waking up and warming up aren't as simple as just throwing on a blanket. According to Bradford's research, warming up is step one. Then you have to halt the use of sedatives as well as the use of the antibiotics and anticoagulation drugs that have been sustaining you. Unfortunately, you won't awaken well rested to a wholesome, warm meal to comfort your return to consciousness. Instead, you'll likely need to continue on a liquid diet for a while, and you'll feel pretty sleep deprived. "When animals hibernate, they're not really sleeping. They will come out of hibernation to sleep," explains Bradford.

Human hibernation itself comes with a myriad of risks. First, when you slow your body down, you also reduce its natural ability to take care of itself, making common infections life threatening. Heck, many aspects of trying to hack the human body into hibernation carry potentially fatal or life-altering dangers—especially considering that research hasn't yet determined the effects hibernation can have on cognition and body functions. Much of the optimism about making human hibernation a reality comes from extrapolating how well naturally hibernating animals like bears, bats, and box turtles can pull it off. Among primates, only one is known to hibernate: the fat-tailed dwarf lemur of Madagascar, which weighs in at about half a pound. Evaluating whether human hibernation can ever work is a question not only of *if* but also of *should*. We can barely comprehend how to make it work beyond an engineering

117

model for a trip to Mars, much less between distant stars.

During my expedition to Antarctica, I studied tardigrades, microscopic animals also known as "water bears" that have become famous for surviving in extreme environments, including the vacuum of space. I filmed the little creatures through my microscope as they woke up from being embedded inside a massive glacier. After slowly perking up, they seemed perfectly happy. Their form of hibernation involves leeching all the water from their bodies when they sense environmental changes and using specially evolved proteins to protect the structure of their cells from desiccation and radiation. In this state they can survive for several years before reanimating when conditions improve. In fact, it seems that they have truly cracked the riddle of hypersleep. But even tardigrades don't have a 100 percent success rate; in my observations, a significant percentage of them simply never revived. Wherever you find inspiration in nature that feels like magic, it's important to remember that nature is not a skeleton key that unlocks all challenges.

Still, the notion of snoozing on long journeys between the stars captures our imagination, perhaps because it gives us the feeling of agency over time. If traveling faster than light is impossible, exploring the possibility of surviving for longer than our natural lifespan grants us a sense of control. Besides, not all depictions of hibernation in space are doom and gloom; they can even offer moments that bring people together. In the TV series *Foundation*, a mother and daughter who were long separated by space and time are able to meet one another far in the future due to hibernation. In *Oblivion*, a spouse is reunited with their partner after being jettisoned out of a space disaster. I'd argue that even in *Alien*, the portrayal of Ripley going into hypersleep with her cat by her side gives me the warm fuzzies. Maybe I can experience the future with my fluffy companion.

Bradford says about his company, "As an advanced concepts organization, we're always looking ten or twenty years into the future and asking how we can improve what we have now. Science fiction is a big motivator for all that." Perhaps science fiction can play a role in imagining more possibilities for how we can safely sleep our way between the stars.

18

CYBORGS

WHAT HAPPENS TO OUR HUMANITY AS WE MODIFY OURSELVES FOR SPACE TRAVEL?

Cyborgs and androids invite both excitement and fear about the future. Uploading brains, overcoming radiation risks, immortality, emotional control, and greater mental capacity are just a few of the opportunities, as presented in science fiction, that could prove useful in space exploration. With each of those prospects, though, comes difficult questions: What do we have to lose? What do we have to gain? And is there a line demarcating what makes us human?

In 2012 I was summoned to serve on a congressionally requested National Academy of Sciences committee to investigate the future of human spaceflight. After two years of research, testimonies, and inspections, we produced a report titled *Pathways to Exploration* that looked out to the 2050s. At the very start of the 258-page report, we included the following caveat:

> The committee...acknowledges the possibility that over the half-century considered, advances in science and technology in bioengineering, artificial intelligence, and other fields may come far more quickly and unpredictably than the advances contemplated for the human spaceflight pathways proposed in this report. Breakthroughs in these other realms could serve to solve many of the large obstacles to exploration beyond LEO [low-Earth orbit]. In particular, the line between the human and the robotic may be blurred more profoundly than simple

linear extrapolations predict. In such an eventuality, exploration of the "last frontier" of space might well occur in a more rapid and far-reaching way than is envisioned in this report; indeed, whether it would still be accurately described as *human* exploration is unknowable.

Doctor Who often probed at the question "human or not?" through its infamous evil cyborgs, known as the Cybermen. When the original show introduced the Cybermen in 1966, they were humans that were upgrading themselves technologically. In 2006 the narrative shifted a bit; the Cybermen were forcefully converting people to cyborgs by putting their brains in robots. Throughout the show's run the Cybermen were generally presented as malicious, barely human beings, until the Twelfth Doctor had an epiphany: the Cybermen were a never-ending emergent parallel to the evolution of human and human-like societies.

Similarly, *Doctor Who*'s Daleks—which are not human cyborgs but rather hybrids of machines and octopus-like aliens—are the classic evil cyborg race. It wasn't until the Eleventh Doctor's companion, Clara Oswald, was unwittingly converted into a Dalek that the show asked audiences to consider where humanity begins and ends when it comes to cyborgs.

I turned to a few experts who have been keenly observing this sci-fi landscape to get their take on whether cyborgs and androids present a cautionary tale or an opportunity to learn how to navigate our future should we begin augmenting humans to cope with the harshness of space. Amber Case, a cyborg anthropologist, considers how we are all cyborgs already. Case has served as a research fellow at Mozilla, MIT's Center for Civic Media, and Harvard's Berkman Klein Center for Internet and Society. Norman Chan is the cofounder of Tested, where he explores the intersection of science, popular culture, and emerging technology, showing how we are all makers. Kishore Hari, a community organizer for science, works with Chan as a pop culture and science contributor at Tested.

ARIEL WALDMAN: Let's start with the basics. What is a cyborg?

AMBER CASE: Cyborg anthropology, the study of interaction between humans and technology, has been applied to hospitals and childbirth and to space travel. The interesting thing about it is that most of the time we think of cyborgs as characters out of *The Terminator* or *RoboCop*, and that's kind of a problem.

The word "cyborg" first showed up in a 1960 paper on space travel. It was defined as "any organism to which exogenous components are added for the purpose of adapting to new ambient spaces." And really it just means anybody who puts a spacesuit on is a cyborg. But even more, if you put flippers on and go into the ocean as a scuba diver, or if you put on an oxygen tank so you can climb Mount Everest, you're adding something to your external self to adapt to a new environment—and that's really what humans are doing every day. We're all cyborgs. Physical extensions of yourself, like a hammer as an extension of your fist or a knife as an extension of your teeth, have been the norm for some time, but now we also have mental extensions of ourselves with smartphones and the internet. It's all really based on the idea of extending ourselves and being part of this larger organism.

KISHORE HARI: I'm certainly a victim of the Hollywood depictions of cyborgs: a merge of biology and technological machinery together into something that is...a new kind of a creature. So I often think of cyborgs in the concept of humanoids. Which is not always the way we have to think about cyborgs when it comes to the science behind it.

WALDMAN: Do androids differ in what they are and how they're presented?

NORMAN CHAN: As a kid watching *Star Trek: The Next Generation*, I was really fascinated by the difference between an

123

android and a cyborg. Cybernetic organisms may have some biological tissue, and in *Star Trek: The Next Generation* you have the Borg as the cyborg. Then you had the perfect android, which was Data, an artificial human. And Data represented for the show the character that was finding his humanity. The classic *Star Trek* trope is that the least biologically human character ends up being the most human. You see these forces at play, too, in shows like *Battlestar Galactica* and what it meant for the analogy of Cylons being indistinguishable from humans.

These kinds of philosophical questions are the hallmark of *Star Trek*, which tries to dive into the deeper questions and reconcile relationships. Why did Data want to be human? Was it meaningful for him to seek humanity, or was that something that was just a directive by his creator? And in *Star Trek: Picard*, we see Data wanting to be a mortal being in his conquest to be human. It harkens back to classic science fiction: Isaac Asimov's [novel] *The Positronic Man* and the movie *Bicentennial Man*, based on the same book, about impermanence being such a quintessential human quality. Now, whether that's something synthetics and humanoids should strive for, that's an interesting debate, right? Does the desire to eventually die make you more or less human? That's an interesting conflict.

HARI: There are also humanoids that have no physical or artificial biology that bring up interesting questions. The Emergency Medical Hologram, for instance, from *Star Trek: Voyager*. The EMH's central character has a whole episode about the concept of whether holograms even have sentience and rights and ethics.

WALDMAN: What questions about humanity are being explored when sci-fi depicts taking pieces of humans, or of what makes us human, and putting them into different bodies?

CHAN: I think *Westworld* and shows like *Black Mirror* tap into this idea of putting human brains into a computer or AI and wading into the debate of whether they should be considered human or not. Even *Star Trek* has that classic original series episode "The Ultimate Computer," with Daystrom putting his brain in the computer and the idea of holographic characters having sentience and individuality. These are all topics that *Star Trek* has broached in the past. In a modern-day science fiction context, both *Westworld* and *Black Mirror* tackle these issues with much more nuance and subtlety. But you're not going to solve all these complicated issues of what makes us truly human when we're installed into a different body in forty-two minutes or a season of television, but you'll at least have some beautiful dialogue and prose that go with it.

CASE: I know that there are a lot of people who say they want to upload their brain and no longer worry about having a body. But isn't a lot of what we experience as humans really embodied in embodiment? What happens if you take away the need to eat? Do you replace it with the need to have electrical power? Does that reduce your humanity?

In the film *The Beyond*, we see a person's brain transferred into a cyborg body, and afterward they just seem...plain. As if the humanity had been taken out. I don't know whether that's great or not. Some might immediately conclude it's not great. But they did it in order to safely travel through space. I do think if we go to space, we are going to need some radiation shielding through some mechanisms that we attach to ourselves or some CRISPR gene editing that prevents us from having lots of mutations or bone density loss. Right now, I don't think, unaided, we're really fit for space.

HARI: I actually think this is one of the more believable components in the movie *The Beyond*. People immediately volunteered to become this robot-suit-brain hybrid to explore the frontier of space in an unhindered kind of way, even if it meant having

to learn how to walk again in a new body. That struck me as just totally true, because we see today people wanting to go on one-way missions to Mars.

There's this trope in a lot of science fiction where the augmentation is about making you more human. And I find that kind of boring now because I don't see humans as any sort of object of perfection. I actually really like River Tam from *Firefly* and *Serenity*. She's totally augmented and has grappled with the ramifications of that augmentation a lot. She had human frailty on display alongside her augmentation and was imperfect from start to finish. I think that's where it's interesting.

WALDMAN: Trust seems to be a common problem with cyborgs in film and TV. *Doctor Who* primarily presents Cybermen and Daleks as unwavering evil. Even in *Star Trek*'s more optimistic views of the future, the Borg is a supervillain, and technologically modified human crewmembers often have plots associated with questioning their loyalty. Should we think of cyborgs as one of us or one of them?

HARI: This is something that's very prominent now in real life. We're having a lot of conversations about trust, generally, just human to human, no less human to humanoid or robot. So it's somewhat natural to see immediate distrust emerge in depictions of cyborgs. I think there's always a sense of paranoia that we see in both science fiction and science reality. The unknown is something I think we have to grapple with as we're exploring. Oftentimes we remember some of our great scientific achievements as this sort of heroic moment, but we sort of ignore the paranoia that emerged around it.

You see this come up as a common theme in a lot of space sci-fi. What does it really mean to be human? What does it mean to belong? Or even what does it mean to be sentient, which *Ex Machina* deals with in a really exceptional way. We know from interviews with Patrick Stewart that he was interested in returning as a character for *Star Trek: Picard* specifically because the script focused on a tension between who belongs and who doesn't, and in that he saw parallels with the Brexit debacle, which attracted him to the story.

CHAN: The idea of the misinterpretation, the miscommunication, that is pure *Star Trek*. A miscommunication, or an unintended communication, leads to a conflict that's a parable for wars in our real world—synthetic humanoids being banned because of a fear of rogue attacks in *Star Trek: Picard*, season one, for instance. It challenges our ethics and simultaneously asks us to analyze the politics of the world being depicted so that we might see ourselves and have our own beliefs challenged.

Paranoia is an apt descriptor of the feelings that stories of augmented and artificial humans tap into. Often the backdrop of space exploration in these narratives acts as an opposing force to that fear, providing a scenario that is so exciting, or impossible to achieve, that it forces us to debate if there should be exceptions or limitations to our caution. One day we may conclude that it's easier to enhance ourselves than to build faster or more protective spaceships. Or that sending an embodied artificial intelligence to explore the cosmos is preferable to sending unthinking robotic space probes. These future crossroads may force us to expand our definition of humanity or accept our limitations in service of our ethics. Either way, given our advancements in biotechnology and artificial intelligence, it's not entirely out there to begin thinking deeply about these concepts today.

19

STARSHIPS

HOW WILL WE TRAVEL BETWEEN THE STARS?

Faster-than-light (FTL) travel could arguably be considered the premier trope of space-based science fiction. How else would mere mortals have adventures beyond a handful of planets if there weren't some way to travel distances so expansive you'd have to break known physics to do it? Okay, yes, through suspended animation, but then we'd be in it just for the destination and not the journey. Plus, I'd like to think that constantly freezing myself to sleep would feel tedious real fast. From the warp drive in *Star Trek* to the hyperdrive in *Star Wars*, the FTL drive in *Battlestar Galactica*, the Stargate in *Stargate*, space-folding in *Dune*, and the Infinite Improbability Drive in *The Hitchhiker's Guide to the Galaxy*—the plot device of FTL travel has been used in so many futuristic films and TV series that it is practically a bedrock of the genre. But it was *Star Trek: Discovery* that pushed FTL travel into new territory with the introduction of the spore drive.

Involving no complicated subatomic particle physics or bending of space-time, this method of FTL travel is a squishier, biological way to skirt around physics as we know it. In the show, the spore drive is based on the discovery of a mycelial network that stretches across the multiverse—essentially microscopic space fungi—and can be tapped into to travel vast distances instantaneously. It reminds me of a debunked theory from the 1800s to explain how light could travel across the universe: a hypothesis that space was filled with a fluid substance, known as ether. All right, maybe the spore drive is just as complicated as a system involving particle physics, but it earns a star for delightful creativity. And it is pretty exciting to dream that "astromycologist" could be a real career one day.

Among the faster-than-light explanations that feel somewhat plausible because of the physics or math behind them, we have to talk about the quantum realm, of which we currently have an arguably nascent understanding. What we do know is that it is a realm where subatomic particles can seemingly travel from one place to another through quantum entanglement and quantum tunneling in ways that don't match up to what's possible in the larger world.

Then there are wormholes, which, while theoretically possible, require several space-time-twisting parameters to even be formed, much less to stay open so people can safely travel through them. You might think of that scene in *Interstellar* when astronauts pass through a wormhole, and the main character, Cooper, nervously asks, "The others made it, right?" and his colleague replies, "At least some of them." Of course, you could also try to journey through time by shimmying up next to a black hole for a quick hour and then jettisoning out of its pull to find that years have passed on your home planet.

Despite the many representations in science fiction, there are no realistically viable ways for us to tap into quantum mechanics, wormholes, or black holes for FTL travel utilizing any foreseeable twenty-first-century technology. Nor are there any physics shortcuts we're currently privy to. If we can glean any lesson from the fact that aliens haven't visited Earth yet—or even left us a voicemail—it's that space is *really* big. What does that mean for our dreams of journeying beyond our solar system?

Mars has remained our "horizon goal" for human spaceflight for generations. Landing a human safely on the surface of the red planet and returning them home is expected to take decades and hundreds of billions of dollars, regardless of which country or company does it first. That doesn't mean we can't begin setting the groundwork for building a starship to carry us *beyond* our solar system. Enter the 100 Year Starship project, which despite its name doesn't presuppose that it will take a hundred years to *build* a starship. Rather, the project is dedicated to spending the next 100 years generating research and nurturing technological, social, and political advancements to prepare us to journey to the stars when the necessary radical breakthroughs occur. The project focuses on fostering an atmosphere where those big, radical breakthroughs can be accomplished today and entirely new disciplines can emerge that go on to influence the future. Think of it as the research and development phase before a project begins.

100 Year Starship was established in 2010 under a partnership between DARPA (the Defense Advanced Research Projects Agency) and NASA. The following year, I was invited to keynote the project's inaugural symposia, where DARPA and NASA were looking for a leader to spearhead the effort. The two agencies located their guiding star in NASA astronaut Dr. Mae

Jemison, the first woman of color to go to space and a powerhouse for improving life here on Earth in her roles as a physician, engineer, environmental studies professor, and social scientist. She majored in chemical engineering at Stanford as well as African and African American studies with a focus in linguistics, all while participating in and choreographing a number of dance productions. 100 Year Starship is at the beginning of a century-long endeavor, but it has already proven that an international community of experts across disciplines—physics, medicine, technology, religion, ethics, politics, and others— can provide fertile ground for the future.

One of the people at the forefront of working on building a starship that could launch well before the next century is astrophysicist Professor Philip Lubin. Lubin began work on his starship concept in earnest after attending the 100 Year Starship Symposium where he was able to connect with other professionals who wanted to support his audacious goal. I first met Lubin while I was advising the NASA Innovative Advanced Concepts (NIAC) program and he had just been awarded a grant to work on his concept for the starship's propulsion. Though it wouldn't be crewed and it might be very small, this ship would send back tiny images from its journey to another star. Unlike chemical rockets, which carry their energy source on board, Lubin's starship would use what's called directed energy—a dressed-up name for using light, specifically photons, as a propulsive force. During a demonstration, Lubin referred to the light coming out of a flashlight he was shining on himself: "The light on my hand is not only illuminating my hand; it's actually pushing on my hand. Light carries energy and momentum."

Think of that every time you turn on the lights or stare at your phone: those photons are pushing on you, but their energy is too weak for you to feel anything. Hence the concept of *directed* energy: the use of extremely concentrated light to produce enough energy to push a starship through space. "The answer is not to make a spacecraft out of a flashlight," Lubin said. "The answer is this: take the flashlight and put it somewhere on the Earth, in orbit, or on the Moon, and then shine it on a reflector, which propels the reflector to speeds that can approach the speed of light. This is a lot like sailing on the ocean: you're pushed by the wind, and the wind then drives the sail forward through the water."

The "flashlight" would be in the form of synchronized lasers erected across an area the size of a city. That much space dedicated to lasers sounds like a lot, but what it could achieve is incredible. "It can push a spacecraft the size of your hand to speeds which are roughly 25 percent the speed of light. That would enable us to get to the nearest star, Proxima Centauri, which is a little over four light years away, in less than twenty years."

Lubin estimates that we could launch a small ship in roughly thirty years, assuming the funding is available. "While it's not going to happen tomorrow, we've already begun the process, and so far it's looking good," he says. "This is both a revolutionary program, in terms of being a transformative technology, but it's also an evolutionary program. So, personally, I do not expect to be around when the first relativistic flight happens, but what inspires me is to look at the ability to achieve the final goal." The thought of increasing our range of space exploration from twenty-four billion kilometers away, or just beyond the edge of our solar system, where the *Voyager 1* spacecraft is currently located, to four light years away is incredibly exciting. *Voyager 1* took nearly half a century to travel that distance. Four light years is thirty-eight *trillion* kilometers, and Lubin is estimating we can get there in only twenty years? Incredible.

So it may be that the first starship to travel outside our solar system will easily fit into a purse. While it isn't exactly the stuff of sci-fi shows, the prospect of humans visiting exoplanets, even robotically, would represent a drastic shift in how far we can go as a species. "Perhaps there's life on an exoplanet, and we would be able to see evidence of that life, either through atmospheric biosignatures or through a dramatic picture," Lubin says. "The consequences are incredibly transformative."

Acknowledgments

I'd like to send a huge geeky hug to the following folks: my friends, family, and colleagues, and especially my husband, Matt Biddulph. Norman Chan, who cohosted, championed, and helped bring my show *Offworld*—about the intersection of space exploration and pop culture, and to which this book owes its origin story—to life. Thanks to my literary agent, Stephanie Kim, and my editor, Britny Brooks, for geeking out with me about this book since its conception. Thank you to Dr. Mae C. Jemison for providing the Foreword and being a source of inspiration. Thanks to Adam Savage and the Tested family, including Kishore Hari, who cohosted a few episodes, for cheerleading, producing, and welcoming *Offworld* as a show on the channel. Thank you to my friends Wes Swingley and Morgan Holzer for being beta testers in reviewing the book. Also, a shout out to two institutions that help push the state of sci-fi and science forward: the Science & Entertainment Exchange, for which I am a consultant, and the NASA Innovative Advanced Concepts (NIAC) program, whose advisory council I formerly chaired.

And a huge heartfelt thank you to every single guest that appeared on *Offworld* and filled my brain with so much excitement: Fatima Abdurrahman, Megan Ansdell, Jun Axup, Gibor Basri, Violet Blue, Penny Boston, Teon Brooks, Bonnie Burton, Michael Busch, Emily Calandrelli, Amber Case, Anthony "Tony" Colaprete, Nicholas de Monchaux, Trace Dominguez, Frank Drake, Camille Eddy, Kimberly Ennico Smith, Lauren Esposito, Rose Eveleth, Nick Farmer, Bobak Ferdowsi, Brian Fisher, Terry Fong, Simone Giertz, Hank Greely, Jasmine Lawrence, Scott Manley, Franck Marchis, Mika McKinnon, Ryan Nagata, Sarafina Nance, Annalee Newitz, James "Jim" Newman, Christopher Noessel, Frances "Poppy" Northcutt, David Pescovitz, Megan Prelinger, Lynn Rothschild, Seth Shostak, Alessondra "Sondy" Springmann, Shannon Stirone, Jill Tarter, Christianna Taylor, Douglas Vakoch, Vicky Vásquez, Indre Viskontas, and Laura Welcher.

About the Author

ARIEL WALDMAN creates imaginative projects that explore the world below our feet and beyond our atmosphere.

She is the author of *What's It Like in Space?: Stories from Astronauts Who've Been There* and hosted and produced the show *Offworld* that explored all things space exploration and pop culture. As an Antarctic explorer, Ariel embarks on expeditions to film life under the ice, which became the subject of her TED Talk. She is a National Geographic Explorer and a consultant to the Science & Entertainment Exchange.

Previously, Ariel has served as the advisory chair to NASA's Innovative Advanced Concepts program and the coauthor of a congressionally requested National Academy of Sciences' report on the future of human spaceflight. An art school graduate who pivoted to science, she was recognized by the Obama White House as a Champion of Change in citizen science. When not in Antarctica, she lives in San Francisco.

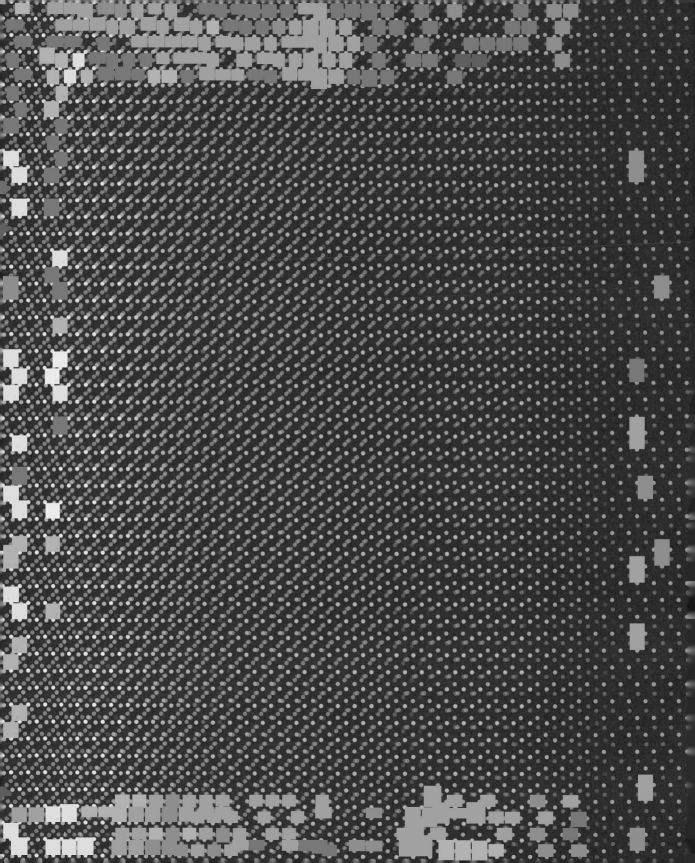